ROUTLEDGE LIBRARY EDITIONS:
THE VICTORIAN WORLD

Volume 13

CONQUEST OF MIND

CONQUEST OF MIND
Phrenology and Victorian Social Thought

DAVID DE GIUSTINO

LONDON AND NEW YORK

First published in 1975 by Croom Helm Ltd

This edition first published in 2016
by Routledge
2 Park Square, Milton Park, Abingdon, Oxon OX14 4RN

and by Routledge
711 Third Avenue, New York, NY 10017

Routledge is an imprint of the Taylor & Francis Group, an informa business

© 1975 Croom Helm Ltd

All rights reserved. No part of this book may be reprinted or reproduced or utilised in any form or by any electronic, mechanical, or other means, now known or hereafter invented, including photocopying and recording, or in any information storage or retrieval system, without permission in writing from the publishers.

Trademark notice: Product or corporate names may be trademarks or registered trademarks, and are used only for identification and explanation without intent to infringe.

British Library Cataloguing in Publication Data
A catalogue record for this book is available from the British Library

ISBN: 978-1-138-66565-1 (Set)
ISBN: 978-1-315-61965-1 (Set) (ebk)
ISBN: 978-1-138-64684-1 (Volume 13) (hbk)
ISBN: 978-1-138-64686-5 (Volume 13) (pbk)
ISBN: 978-1-315-62734-2 (Volume 13) (ebk)

Publisher's Note
The publisher has gone to great lengths to ensure the quality of this reprint but points out that some imperfections in the original copies may be apparent.

Disclaimer
The publisher has made every effort to trace copyright holders and would welcome correspondence from those they have been unable to trace.

CONQUEST OF MIND

Phrenology and Victorian Social Thought

David de Giustino

CROOM HELM LONDON

First published 1975
©Croom Helm Ltd 1975

Croom Helm Ltd
2/10 St. John's Road, London SW11

CONTENTS

Acknowledgement		vii
Abbreviations		viii
I	Reconnaissance	1
II	A Science of the Mind	12
III	A Respectable Concern	32
IV	A Shortcut to Knowledge	58
V	Transmission and Schism	77
VI	The Ungodly Error	104
VII	The Remaking of Man	136
VIII	The Philosophy of Education	165
IX	The Politics of Education	196
Afterword		227
Bibliography		230
Index		244

To my parents

Acknowledgement

A modern study of phrenology and Victorian social thought should have appeared years ago and even this insufficient volume was unduly long in the making. Of course there has been some warning of its coming, and I am grateful to the editors of *Victorian Studies* for their mention of the project and for their permission to publish once again the material contained in Chapter VII. If any of the leading phrenologists of the last century could now read the following text, he might well issue warnings of his own about any errors or distortions (for which I must confess to being solely responsible); but he would also marvel at the uncommon 'Benevolence' of all those who have helped me in this work. I am particularly thankful to Professor John Harrison of the University of Sussex, Dr F. B. Smith of the Australian National University, Mr Russell Blair of Honolulu, Mr Roger Cooter of Churchill College, Cambridge and Professor A. M. McBriar of Monash University Melbourne for their advice, support and optimism. Above all I shall always appreciate the fellowship I held in 1973 at the University of Auckland in New Zealand. It proved a most agreeable place in which to teach and write as well as to gain a certain perspective of Victorian history.

Lexington, Michigan
August 1974.

Abbreviations

Journal (or *Phrenological Journal*): the *Phrenological Journal and Miscellany*
NLS: the National Library of Scotland, Edinburgh
B.M.: the British Museum, Additional Manuscripts
C.P.: the Richard Cobden Papers, British Museum
n.p.: no place of publication cited
n.d.: no date of publication cited

Chapter 1 Reconnaissance

It seems that Coleridge was once given several pamphlets dating from the reign of Charles I. After reading the material, he decided that most of the conclusions were sensible although the underlying assumptions were often wrong. How very different, he thought, was this Jacobean writing compared to that of modern times, when 'premises are commonly sound, but the conclusions false'. Perhaps this was a hasty judgment, rendered by a man who admitted that he read only to amuse himself. He should have allowed for at least one exception to the rule, for what he had to say about the logic of seventeenth-century polemics he might also have said about a new scientific philosophy in his own generation.

Early in the nineteenth century many people accepted both the assumptions and the conclusions of phrenology. The science of 'head-reading', as it was sometimes called, was widely received as the latest and most significant step towards a complete psychology of man. Today the temptation exists to remember phrenology as a clever man's living at the county fair, the embarrassing pastime of eminent Victorians or, at best, a primitive variety of psychoanalysis. It is unfortunate that this should be the case because for a time phrenology proved useful to thousands of believers. They were not conscious of defending a 'fringe science', nor was it possible for them to know that they were responsible for 'psychology's great *faux pas*'.[1] On the contrary: the dedicated phrenologist of the nineteenth century was confident that he possessed a truly scientific understanding of human nature. Even Coleridge, for all his concern about sound logic, thought that phrenology was 'worth some consideration' and that 'all the coincidences which have been observed [in phrenological analysis] could scarcely be by accident'.[2]

Logical or not, a scientific theory of the mind was long in coming. The obstacles were, and continued to be, formidable. The lack of anatomical knowledge meant that it was necessary to approach the mind philosophically: its operation was usually explained in the language of metaphysics rather than in terms of anatomy or physical science. Psychology (if we can call it that) was in the hands of men who were either unable or unwilling to come to grips with the relationship between the mind and the brain – a relationship which, at the beginning of the nineteenth century, surprisingly few people took for granted. Psychologists were still impressed by, and satisfied with, the picture of the mind which they had historically achieved through philosophical analogy and pure logic.[3] Physiology had little importance to the study of human behaviour and the physiology of the brain had even less. Given these conditions, the work of the Viennese physician Franz Anton Mesmer in the 1780s and nineties was revolutionary: he attempted to show that a state of mind was obtainable by artificial means and could be used to alter or influence human conduct. In effect Dr Mesmer and his followers thought it possible to conquer the mind through knowledge and then, using the same anatomical knowledge, to subdue certain mental impressions or introduce others. Of course not everyone wanted to accomplish quite as much. It was enough to understand the mind; the idea of controlling it, for the purpose of reducing pain during surgical operations or for any other reason, remained an utter fancy. But thanks to the work of Mesmer and other theorists at the end of the eighteenth century, the scientific atmosphere was greatly changed as the mind was increasingly linked to the brain and to the function of the nervous system. Certainly the research was imperfect, but it is true that the mystery of the mind was considerably unravelled and that contemporaries were conscious of the progress. Sir Charles Bell, for example, thought his own research ranked in importance with that of Harvey and Newton. In 1811 Bell published the *New Idea of the Anatomy of the Brain*, in which he distinguished between sensory and motor nerves and spoke of the brain as a clearing house for the impressions received from 'bundles of different nerves'. While it was almost another twenty years before Bell's findings were truly appreciated, the course of physiological psychology was now set. The over-riding effect of the generation which began with Mesmer and ended with Bell was to remove the study of the mind still further (and perhaps decisively) from the world of philosophy and logic.

Even so, psychology was to remain in a state of flux. The debt to pure logic died hard with men like James Mill, his son and the society for whom they wrote. The age of Coleridge and Mill did not rule out the possibility of understanding the mind anatomically, but it did cling happily to the smug speculations of the past. The mind was likened to a vault of good and evil forces, a repository whose precise dimensions and whose riches varied from one person and nation to another. It was an age already disposed to make the most remarkable distinctions about intellectual ability and character. The most awesome judgments were delivered with an almost careless grace. 'It is to the cultivation of the moral qualities', asserted the *Morning Chronicle* in 1815, 'that England is indebted for her power and influence. From the want of them, France may be mischievous, but she will never be great.'[4] At the end of a long struggle with revolutionary France, the *Chronicle's* arrogance is probably understandable; still, the fact remains that people were accustomed to far-reaching estimates of the mental powers of others. They were also receptive to a theory of human behaviour which allowed them to make such estimates: a theory which incorporated historic, common-sense conclusions and the recent discoveries of science.

The ideas put forward by Drs Gall and Spurzheim at the very beginning of the century seemed to provide the appropriate mixture. Phrenology was born of the same interest in anatomical research which inspired Charles Bell while it also provided a new scientific certitude for a host of older notions about human behaviour and society. In producing this popular mixture, phrenologists satisfied their own craving for a philosophy of man and at the same time met the public need for a popular science. For several decades phrenology was anything but an apparent failure. It was a complete science by itself, and books on phrenology found a wider appeal than many sound essays whose value has never been questioned. It took more than fifteen years for the *Origin of the Species* to sell only 16,000 copies in England; the *Constitution of Man* (the most popular book on phrenology) sold 2000 copies in ten days in 1835 and was soon found in respectable libraries and mechanics' institutes all over Britain. Its author, once praised in the *Imperial Dictionary of Universal Biography* as one of the 'few thinkers who have exercised so wide an influence in the present age',[5] was forgotten along with the pseudo-science which he always defended. His career — in almost every respect a personal success story — remains central to the history of phrenology in Britain.

George Combe was born in 1788. His home in the Livingston's Yards of Edinburgh was located in a damp place beneath the steep bluffs which support the Castle. The house of his father, a brewer, was much too small for the seventeen children who eventually lived there; the damp conditions contributed to the poor health of George and his brother Andrew and to the death of several of the other children. For his education George was sent to the parish school and then endured a harsh training at the Edinburgh High School. There was nothing very unusual about his religious instruction; it was relentless. On Sunday the brewery gate was closed, the children rose at eight and went to the

West Church until one. After lunch they returned to the church until four o'clock, and after supper retired to the drawing room where they read psalms, memorised the *Shorter Catechism* and each read a portion of the Bible before being sent to bed.

In this rather grim atmosphere, George Combe was uncertain about his career. He studied for a time at the University of Edinburgh where he thought he might like the life of a political journalist; he joined the Cobbett Club and faithfully read the *Political Register*. Without taking a degree from the University, he secured temporary employment in an Edinburgh law firm. He felt so much at home in this business that he underwent training, was admitted as a writer to the signet in 1812, and then set up practice for himself. For Combe these were hectic years spent, as he later recalled, in self-improvement. In order to perfect his style of writing and speaking, he took lessons from a private tutor during his supper-hour and then hurried back to the law office. He kept himself to a schedule which (so he believed) gave him the desired confidence in public speaking as well as a head of grey hair and chronic indigestion. To his nephew Robert Cox, George Combe gave the advice typical of a self-made man: always work hard and never think of postponing or giving up a task.[6] The Combe brothers were good at taking their own advice, for neither of them was afraid of hard work. Andrew Combe, born in 1797, was even less decided about a career than George and was far less successful in school; but he applied himself to his studies, overcame the handicap of extremely poor health, and became one of Scotland's most distinguished physicians. Fame and fortune came so quickly upon Andrew that when his health finally forced him to retire early from medicine he could still easily afford winter trips to the Mediterranean and on one occasion thought nothing of contributing 200 guineas to support the Phrenological Society of Edinburgh.

The financial success of the Combes brought them a comfortable distance from the ancestral home in Livingston's Yards. George was particularly careful in money matters. Despite the duties of his law practice and the demands made upon him as a writer and lecturer, he continued (with the help of his brother William) to manage the family brewery for many years after his father's death. When he saw that William, a brewer by trade, was determined to move to New York, George sold the old brewery to city commissioners for the sum of £9650, which was divided among the brothers. Tempted by boom conditions in America, George then put much of his fortune into New York and Illinois railroads, and by the 1840s (notwithstanding his losses during the 1837 panic) he derived over £200 annually from these investments. Combe's supposed love of money caused chatter among his critics, who joked that he had chosen his wife, Cecilia Siddons, not only for her phrenological endowments but because she brought him £800 yearly.[7] And when he made his phrenological speaking tour of the United States in 1838–40, Combe encountered the complaint that he had come to make money, which he thought 'a vulgar charge'. He did not understand why 'such a people, who are always talking of dollars and percents and charge me high for food and drink, should reckon it any disparagement in me to desire the natural fruits of my talents'.[8] Combe's self-defence was entirely sincere; in the closest examination of conscience he did not believe himself a mercenary Scot, nor did any of his friends and acquaintances. In her *Biographical Sketches* (in which Combe and Von Humboldt were the only two men under the heading 'scientific'), Harriet Martineau remarked that George Combe's 'intellectual and spiritual gifts' were unfortunately of low quality, but at least he stood 'above the perils of gross selfishness and dishonesty' and treated others justly and benevolently.[9]

Wealth was, after all, necessary for comfortable and respectable living, and Combe disliked those who accumulated coin for coin's sake. The religious values of his youth persisted: a person's wealth was a factor in judging his character, although Combe insisted that he wanted nothing to do with the 'money-worshippers' whom he found to inhabit New York — that 'commonwealth of mammon'.[10] Money provided everything that was important to Combe: travel, education, and the opportunity to be a philosopher. George had his first glimpse of the good life at the age of eleven, when his father took him and his brothers John and Abram to see a performance of 'As You Like It'. The drama, the costuming and the scenery all opened to Combe a view of social refinement which he had never before experienced; the only concept which his evangelical mind thought comparable to this theatrical production was that of 'paradise before the fall'.[11] It was then, according to Combe himself, that he resolved to make the crisp and clear language of gentlemen his own and to strive for the refinements and comforts of a large townhouse. He succeeded to the extent that George Eliot later 'shuddered at the impiety' of anyone who doubted that Combe's elegant house in Melville Street was *'le meilleur des mondes possibles'*.[12]

Combe had his own role to play on the stage of Edinburgh society. His profession of phrenology may have seemed rather silly to some people, but as a successful lawyer and writer he was not easily overlooked when invitation lists were drawn up for Edinburgh dinners and receptions. He was even present at the reception given for Francis Jeffrey when he became Lord Provost in 1832, notwithstanding their long feud over the merits of phrenology. Combe's world reputation as a social reformer gave him entry to the best philanthropic circles in which he moved, perhaps surprisingly, with an assured poise. In America his friends included Samuel Gridley Howe, William Ellery Channing, George Bancroft, and Nicholas Biddle; Horace Mann was so impressed by Combe that he named a son after him. Rembrandt Peale painted Combe's portrait 'for his own gratification' and Americans so appreciated the honest and somewhat blunt Scot that they elected him to the National Academy of Natural Sciences and to the American Philosophical Society.[13] At home the political atmosphere of Edinburgh was decidedly whiggish, and when Combe applied for the Chair of Logic at Edinburgh University he reminded certain officials that he had always supported the whig cause.[14] Combe was so confident of his political connections that he once imagined he might turn a law career into a political one, but he felt that the 'more aristocratic circles in Scotland' blocked the way.[15] Essentially Combe considered himself a middle class liberal in politics and it was among British liberals that he found most of his friends and well-wishers.

Perhaps Combe's American friends were more fortunate for knowing and admiring him from a distance. The demands were stiffer upon those who knew Combe at closer range. As Harriet Martineau remembered him, Combe was not financially selfish but he was mentally stingy. He never closely examined the opinions he so readily rejected, nor was he known to be very respectful of the attainments of others — unless, as Miss Martineau remarked, they 'were offered to him as confirmation of his own philosophy'.[16] Some of the opinions which Combe was quickest to ignore belonged to his old friend Hewett Watson, the botanist who was among the precursors of Darwin. Watson had the habit of being terribly frank. 'Let us admit it', he wrote to Combe in 1840, 'you are excessively tenacious of ideas once entertained, and even if they happen to be erroneous, it is exceedingly difficult to get them supplanted.'[17] Watson was annoyed by Combe's egotism which, while not always apparent, was never far below the surface. When touring Germany in 1842, he tried to take stock of his position among the great British minds of the day. He concluded that his own 'solidity and consistency of observation and reasoning' warranted him a place 'equal to that occupied by men who are regarded in Britain as standing far higher than I am allowed to do'.[18] British phrenologists realised how much Combe craved attention and deference; they heard him say that he was 'like other prophets who have no honour in their own land'. Certainly when he visited London in 1824 Combe expected all the acclaim due a prophet; instead he was dissatisfied at having to take many of his meals alone and conducting phrenology seminars much smaller than those he knew in Scotland. To make amends for their neglect of him, the Phrenological Society of London bestowed honorary membership upon Combe — a ploy which contented him only briefly.

While most self-made men are given to conceit by degrees, in Combe's case the process was unduly rapid. He kept to his own schedule and to his own opinions, and he prided himself for his independence of mind and fortune. He did not like to be kept waiting, and he published the tardiness of anyone with whom he did business. If punctuality was not one of the departments of the brain, it might as well have been as far as Combe was concerned. He often boasted of the time-table which he kept. The virtuous life was one in which a man had plenty to do and did it on time. Frances Kemble, a cousin of Combe's wife, noted that during his American tour 'Mr Combe had parceled out both his whereabouts and whatabouts, to the very inch and minute, for every day in the next two years to come.'[19] The reason for this fussy schedule was simply that Combe believed reports (sent by Miss Kemble and others) that everyone in New England was reading the *Constitution of Man*, and he wanted to accommodate as many of his admirers and fellow phrenologists as he possibly could.

Having gained national attention for his often quarrelsome defence of the 'new philosophy' of phrenology (of which he was the undoubted leader after the publication of the *Constitution of Man*) George Combe offered himself as candidate for the Chair of Logic at Edinburgh University. His approach to this position was typical of his feelings about phrenology and his role as the leading phrenologist. Since the Modern Athens was pre-eminent for its study of philosophy in the north, was it not only proper for Edinburgh in 1836 to be the first to honour the most modern branch of philosophy by giving its leader an academic chair? Combe looked upon a university post as an effective way of advancing phrenology itself while at the same time adding to Edinburgh's 'enlightened sagacity';[20] he also thought that in competition with such career philosophers as Sir William Hamilton ('that representative of the old metaphysics') he had a good chance of winning. When the votes were tallied Combe came last in a list of four and Hamilton won easily. After this experience Combe thanked all those who had written testimonials on his behalf and confessed that he had not expected to win anyway. His consolation was the number and variety of persons who (in spite of all his reputed *self-esteem*) publicly supported his candidacy. Naturally there were several medical men, including Dr Carmichael in Dublin and Dr Otto in Copenhagen; there were also a few non-phrenologists who 'perceived a lot of good in phrenology', such as Dr James Johnson, principal physician to King William.[21]

While it ended in disappointment, Combe's application for the university chair was not really a startling affair. Other phrenologists applied for chairs, usually in the sciences, and several of them were successful. Combe's loss in Edinburgh did not mean that universities were out of bounds for phrenologists. Many phrenology lectures and debates were already held in university halls and in 1837, the year after his Edinburgh application, Combe received a letter from America inviting him to accept the chair of 'mental and moral philosophy' to be established at one of the new state universities. The letter, sent by a Scot who was a graduate of Glasgow University and now a regent of the American institution, begged Combe to consider lending his 'talents, reputation, and best efforts for the reknown and prosperity' of the University of Michigan.[22] Combe thought little of the offer. True, the salary was enticing (professors were to be paid more than the state governor, who got $2000 per annum), but Combe disliked the idea of frontier life and could not believe the regent who insisted that the climate of Ann Arbor was as healthy as any part of the world.[23] Eventually Combe replied that were he only twenty-five he might accept, but he preferred to 'dedicate his remaining years of health and strength' to the fields he knew best; indeed, this was his duty.[24]

Combe's sense of duty saw him admirably through life and helped him to achieve a fortune and a philosophy. The one came almost as quickly as the other. Perhaps because they had themselves risen from the shadows of Livingston's Yards, the Combes — Abram, George and Andrew — viewed man's progress with a certain optimism. Their family background and their education made them products of a preindustrial Britain, and their concept of man as a rational being derived almost completely from the philosophy of the last century. They were not modern figures with an unbounded trust in the future; they looked upon the growing industrial cities as uncivilised places divided 'between methodists, socialists, and radicals ... all buried in smoke, ignorance and dirt'.[25] Where was one to find any order or a pattern by which man might manage his affairs? William Combe, who moved to America, was not interested in such questions; Abram Combe, who followed Owen, discovered a social philosophy but lost his money and his health in the process; that left George and Andrew, whose professional training inclined them to a reverence for the 'fixed laws of Nature'. George Combe's reading of Malthus in 1805 came as a 'flash of light': he found it comforting to read what he had always accepted, that everything in Nature, including the diverse population of the world, proceeded according to a body of laws.[26]

Combe was almost thirty years old when he discovered a complete mental philosophy, an all-embracing system of moral and physical order. Already he assumed that the sciences of government, education and human conduct all rested on the admission of laws which, as far as he could tell, no one dared to dispute. At first Combe refused to believe that phrenology had anything to do with these Laws of Nature; he was so amused and persuaded by the critical remarks of the Edinburgh *Review* in 1815 that he had no wish to see the curious Dr Spurzheim. Without warning Combe was introduced both to Spurzheim and to phrenology. Invited to a friend's house, Combe found himself a guest along with Spurzheim, who turned the gathering into a phrenology seminar. Combe listened politely as the German doctor reviewed the principles of his science and dissected a brain, which he had brought in a paper bag. The demonstration aroused Combe's curiosity and, 'always eager for knowledge', he began to study anatomy and to attend Spurzheim's remaining lectures. His brother Andrew was equally cautious at first; but nearing the end of his medical training and having some free time, Andrew also investigated phrenology. His attention was drawn to Spurzheim's *Observation sur la phrénologie*, which impressed him as a sensible new psychology.[27] Reluctantly at first, these two Scots, self-made men and dilettantes of philosophy, approached the very answers they first received with derision. In small doses they imbibed a new science of man.

In the years following George Combe's death in 1858, few persons hurried to wring from history a judgment of his career and his talents. Already it seemed safe to say (and Harriet Martineau said it) that George Combe deserved no special honours as a genius, a philosopher, or an original thinker.[28] In 1869 this was to judge Combe by the short memory of phrenology's decline; it was also to neglect Combe's role in making phrenology attractive to his contemporaries. When Combe declared in the preface of the *Constitution of Man* that phrenology was the 'clearest, the most complete, and the best-supported system of human nature',[29] there were many who agreed with him. The readers of the *Constitution of Man* did not dwell on Combe's supposed selfishness, his middle-class opinions, his quest for fame, or even his 'mental stinginess'; they noted instead his observation that no-one had ever demonstrated the relationship of the Laws of Nature to the total character of man, and they were anxious to see whether phrenology had an answer. Why should Combe's theory of the mind have found a much broader appeal than the ideas of Charles Bell? Why did this pseudo-science appeal to readers of all classes? Phrenology was a world-wide phenomenon and counted believers everywhere; but before looking at the devotion of the faithful, it might be wise to investigate the origin and progress of their creed.

Notes

1. J. C. Flugel, *A Hundred Years of Psychology, 1833–1933; with an additional Part on Developments, 1933–1947* (London, 1951), p.44.
2. *The Table Talk and Omniana of Samuel Taylor Coleridge*, ed. T. Ashe (reprint: London, 1923), 24 June 1827.
3. Flugel, *A Hundred Years*, p.36.
4. *Morning Chronicle*, 2 February 1815.
5. John P. Nichol, in the *Imperial Dictionary of Universal Biography* (Glasgow [1880]), VI, p.1100. Nichol, a phrenologist himself, was professor of astronomy at the University of Glasgow.
6. Combe to Cox, 16 March 1837. NLS 7387/339.
7. Joseph MacCabe, *Life and Letters of George Jacob Holyoake* (2 vols., London, 1908), I, p.27.
8. Combe to Dr Brigham, 15 October 1838. NLS 7395/52.
9. Harriet Martineau, *Biographical Sketches* (London, 1869), p.274.
10. Combe to Horace Mann, 20 January 1839. NLS 7396/17.
11. Charles Gibbon, *The Life of George Combe* (2 vols., London, 1878), I. p.56.
12. *George Eliot's Life as related in her Letters and Journals*, ed. J. W. Cross (3 vols., London, 1885), I, p.292.
13. John D. Davies, *Phrenology: Fad and Science. A Nineteenth Century American Crusade* (New Haven, 1955), p.21f.
14. Combe to John Conningham (Solicitor General), 10 May 1836. NLS 7386/559.
15. George to Andrew Combe, 9 November 1845. NLS 7390/207.
16. Martineau, *Biographical Sketches*, p.271.
17. Watson to Combe, 17 April 1840. NLS 7257/112.
18. George to Andrew Combe, 10 May 1842. NLS 7379/86.
19. Frances Kemble, *Records of Later Life* (3 vols., London, 1882), I p.167.
20. Combe to Jeffrey, 9 April 1836. NLS 7386/532.

21. A seventy-nine page booklet, *Testimonials on Behalf of George Combe, as a Candidate for the Chair of Logic in the University of Edinburgh* (Edinburgh, 1836) gives the statements of about forty persons as well as a defence of phrenology by Combe.
22. Adam to Combe, 12 August 1837. NLS 7242/7.
23. *Ibid.*
24. Combe to Adam, 13 October 1837. NLS 7387/431.
25. Andrew to George Combe, 12 April 1840. NLS 7254/32.
26. George Combe, *The Relation between Science and Religion* (Edinburgh, 1857), xii.
27. George Combe, *The Life of Andrew Combe, M.D.* (Edinburgh, 1850), p.66.
28. Martineau, *Biographical Sketches*, p.276.
29. Combe, *Constitution of Man* (1st ed.: Edinburgh, 1828), viii.

Chapter II A Science of the Mind

In the very month of Waterloo, the Edinburgh *Review* was distressed to find that Britain had been invaded from the continent. True, it was not the sort of invasion which many people had feared for years (and which just as many had long ago discounted), but to some observers it was almost as alarming as the prospect of French troops in Kent. This was the attitude of one contributor to the *Review* who deplored the spread of 'continental quackery' in Britain. In his opinion, the easy access which 'foreign imposters' had to Britain and their success in deluding so many people was a national disgrace. In effect the whole country had become 'a field for quacks to fatten in', and the writer (who was probably a Scot) felt that England had already become 'the sweetest corner' of that pasture.[1] The writer had a particular form of quackery in mind. He was referring to the ideas of Johann Caspar Spurzheim, who had been lecturing in the country for some time. Partly philosophical and partly physiological, Spurzheim's message went under various names. Some described it as cranioscopy; others called it organology; perhaps the most accurate (and least mysterious) description was 'cerebral anatomy'. After 1815 these expressions gave way to 'phrenology', a word which Spurzheim himself preferred to all the others, and one which the Edinburgh *Review* never failed to use with contempt.

Whatever they thought about Spurzheim's invasion and his recruitment of collaborators, the Edinburgh *Review* at least understood something of the history of phrenology. Like most critics of the new science, they knew that Spurzheim alone was not responsible for 'cerebral anatomy'; they realised that phrenology was the product of Spurzheim's long association with his friend and mentor, Dr Gall. There is in the history of ideas a certain risk in speaking of anyone as an originator, but in so far as the word had any meaning to the Edinburgh *Review* (or to Spurzheim, for that matter), the originator of phrenology was indeed Franz-Josef Gall. Born in Baden in 1758, Gall started his medical studies in Strasbourg and finished in Vienna, where he soon enjoyed a successful practice. His fame was such that, while still quite young, he was invited to become one of the Emperor's physicians. He declined the offer not from a sense of duty to his present patrons but because he wanted more time to resume the enquiries he had begun as a student. While at university Gall had often wondered why students who excelled in languages or had good memories also seemed (to him, at least) to have 'large prominent eyes'.[2] The tentative explanation was that some relationship existed between the facial features and character — between the shape of the head and the powers of the intellect. Gall's research in the late 1790s appeared to confirm this theory; as a visitor to the hospitals and asylums of Vienna he became convinced that the desparate faces of the inmates fairly reflected their mental afflictions. Perhaps Gall expected a great deal from these observations and could not therefore be disappointed. In any case, the result of his work was pretty much what he had hoped to achieve: a 'perfect knowledge of human nature',[3] a psychology which supposed that the brain was the seat of the mind and the place where all ideas are born. From this it followed that the interior mass of the brain, with all its strengths and weaknesses, caused a corresponding relief on the outside of the head, where, according to Dr Gall, the 'most fundamental qualities and faculties' were visible to all.[4]

These notions were not entirely novel. The ancients had been fond of speculating about the localisation of mental faculties; Galen and Nemesius made the most of physiognomy in their day, and medieval Europe followed their lead. The Platonic analysis of the mind remained a remarkably durable one, as succeeding centuries arranged and rearranged the faculties into groupings; the powers of sensation and imagination were generally located in the front of the brain, those of emotion and memory in the back, and those of reason in the middle. In the eighteenth century, most of these ideas were still current, if not fashionable. At the same time, there was considerable dissatisfaction in the medical profession with the theories of physiognomy and cerebral localisation. The answers now seemed simplistic and facile; above all, they were too speculative for the Age of Reason. These reservations were expressed by a number of writers not long before Gall began his career. In 1784 George Proshaska published his *Dissertation on the Functions of the Nervous System*, in which he conceded that the brain was probably composed of parts all designated by Nature for a particular use; but he did not think it was yet possible to say very much about the essential properties of these faculties. For a while Gall shared these misgivings. Like Proshaska, he wanted to rid psychology of 'abstractions'; he wanted to find definite physical evidence for the operation of the cerebral organs. Ultimately this meant a biological rather than a speculative psychology, for Gall believed that the mind could be studied anatomically.

It was this anatomical, or rather neuro-anatomical, emphasis which linked physiognomy to phrenology and which won for Dr Gall the brief attention he enjoyed as the originator of a new science. The research which he began about 1800 (assisted by his pupil Spurzheim) involved the dissection of hundreds of human and animal brains and the examination of cranial nerves. A modern writer has recently commented that this research was a diversion from Gall's original approach and that his work during these five years neither contradicted nor proved his earlier opinions.[5] Unfortunately it is the only conclusion we can reach. But Gall and Spurzheim felt differently. They admitted only that their research was incomplete: that the mental organs were not all marked off precisely, or that their functions and 'external signs' deserved further study. They granted that there was still much to know, and they were certain that neuro-anatomical investigation would confirm what they already understood about human behaviour. A scientific knowledge of the functions of the brain would show, once and for all, which were the 'fundamental, primitive faculties' of the mind,[6] and would put an end to all the 'abstractions and generalities' which were the bane of physiognomy. The guesswork of the past, which connected character to one's visage and said no more, was now eliminated by Gall's discovery that 'external signs' depended upon the internal geography and operation of the brain.

Of course Dr Gall's proof was not conclusive. It is also true that his work was not very systematic and that he never really explained how the brain functions. But he claimed to be on the right track, and the Austrian Government soon noted his claims. They did not approve. Gall's 'cranioscopy' was regarded as dangerous to traditional morality and religion because it smacked of materialism; eventually he was ordered to stop lecturing. In vain did he defend himself by pointing to the popular physiognomy of the ancients; official displeasure finally caused him in 1805 to leave Vienna. Joined by Spurzheim, Gall embarked on a lecture tour of Europe. The trip proved a remarkable success. In Germany, Holland, and Denmark, respected surgeons and scientists paid their compliments; they were not all convinced of Gall's physiological doctrines, but they were eager to learn of his neuro-anatomical findings. Even at this early stage of its progress, phrenology's connection of psychology to anatomy seemed to impress those outside the world of science much more than it did those within; Goethe, who followed Gall from one lecture to the next, was willing to accept the connection in principle.[7] By the time Gall and Spurzheim reached Paris in 1807, their fame had preceded them, for during their tour they had refined and developed their ideas into a body of medical principles. Compared to their misfortunes in Vienna, their stay in Paris was triumphant and augured well for the new science. Parisian society flocked to Gall as it had to Franklin, and their generosity led to the publication of the *Recherches sur le systeme nerveux en general, et sur celui de cerveux en particulier*, the first part of which appeared in 1809. Gall soon attracted a following among medical men, resident ambassadors and French writers, including Saint-Simon. So much did Gall feel at home in Paris that he politely declined the invitations of the Austrian emperor and Prince Metternich to return to Vienna.

Gall remained in Paris and became a French citizen, but Spurzheim was made of different stuff. Younger and more independent in spirit, he naturally desired a reputation of his own and was willing to resume his travels. His journeys were to take him far from the anatomical phrenology which Gall prescribed, for Spurzheim was interested in applying phrenology to problems of contemporary philosophy, religion, and social reform. Admittedly Gall was concerned with human character and behaviour and occasionally he too indulged in character analysis. But his primary intent after 1800 was relatively simple: he only wished to prove that psychology depended on the structure of the brain, and the reputation which he enjoyed in France and Germany was essentially based on his methods of dissection and anatomical research. In a sense Spurzheim reversed the emphasis. His objective was a whole system of philosophy in which surgical dissection was only the first step. To Spurzheim phrenology implied a self-contained philosophy of the mind, and Gall never forgave him for reviving the 'abstractions and speculations of the philosopher'. By the time Spurzheim crossed the Channel in 1814 to preach on his own, his long association with Dr Gall had virtually come to an end.[8]

The phrenology which Spurzheim brought to London must have seemed a variation of Gall's craniology. The same language was used to identify the organs; the same surgical techniques were used to prove scientific accuracy; the same respect for Nature was professed in order to defend scientific inquiry. Moreover, all phrenologists – whether they learned their lessons from Gall or from Spurzheim – agreed that a mere list of mental powers was not enough. To them Nature suggested the division of these powers into two main categories. The first included those 'feelings and propensities' which were more behavioural tendencies than they were the generators of ideas. Common to man and beast, the feelings were appropriately located at the bottom and the back of the brain and included *adhesiveness* (the tendency to live in communities), *combativeness, destructiveness, acquisitiveness* and *amativeness* (usually described as 'the feeling of physical love').[9] Into this category also fell those faculties called *sentiments*, which caused a particular feeling such as *self-esteem, love of approbation, benevolence* and *cautiousness*. A second order of Sentiments were those proper only to man, such as *veneration* (piety, devotion), *hope, ideality* (inspiration or imagination), *conscientiousness* (duty or gratitude), and *firmness* (resolution or determination). Most of the nobler sentiments were found near the crown of the head.

In the second general division, there occurred the 'Knowing Faculties', by which a person understood the existence of external things and their qualities (*individuality, form, size, weight, colour, locality, order* [arrangement], *time, number* and *tune*. These powers all relied upon the sense perceptions. Also in the second category were the 'Reflecting Faculties' which 'distinguished man in an eminent degree from the lower animals';[10] these comprised *comparison, wit, causality* (the ability to trace cause and effect) and *imitation* (by which faculty children are first taught anything). Spurzheim and his immediate followers in Britain believed that there were probably thirty-three faculties altogether, whereas Gall detected only twenty-seven. The discrepancy was not very upsetting, for Gall had long before admitted that future research might well uncover additional faculties. Both sides also agreed that each of the mental faculties exercised a particular role in human behaviour and that each was subject to the disease of imbalance. For example, *adhesiveness* (a power common to men and animals) was proper when it disposed persons towards friendship and to society in general; but in an excessive or over-developed state this faculty became the victim of 'nostalgia' — the person could not bear the thought of being separated from his friends or from his usual living quarters.[11] Similarly, the faculty of *benevolence*, when operating properly, produced a cheerful and even temper and the desire for the happiness of others, while an excess of the same power led to the foolish profusion of gifts or to a naive trust of strangers.[12]

But it was Gall's apprehension that this sort of psychoanalysis could be carried too far. He particularly resented (in the absence of final medical evidence) Spurzheim's claim that there was an interaction of the mental faculties. If, according to Spurzheim, *veneration* and *hope* were both large while *conscientiousness* and *benevolence* were small, the individual was likely to prefer the pomp and rituals of religious worship to the practice of charity and justice. If in another person *cautiousness* were large compared to a small *combativeness* the individual might be 'extremely timid'.[13] Gall may have wanted to believe in such judgments, but he saw before him the danger of being unscientific and of entering the quicksands of 'speculation'. In the light of what phrenology had to say, however, their dispute was a slight one. Craniologists on both sides of the Channel believed that the mental organs were intrinsically good because they were endowed by Nature; they also believed that an exact knowledge of the brain would eventually classify, if not analyse, all human actions, just as the ancients had classified man's physical appearance by means of the Four Temperaments. Of the two aspects of human existence, the mental constitution was by far the more complex, and only the science of Gall and Spurzheim gave any promise of revealing the nature of mental powers. 'Phrenology is the key to the mind', wrote Andrew Combe to Sir James Clark in 1842. 'It shows where reliance can safely be placed and gives *just* confidence in the good of human nature..'[14]

Once the student of phrenology mastered the location of the faculties and understood something of the structure of the brain, it was not difficult for him to render a fair judgment of a person's mental character. Of course the probability of error always existed and was admitted by the best phrenologists; in addition, there were special risks in analysing persons who had 'passed the middle period of life'. Phrenologists believed that after middle age both the brain and body began to decay, and occasionally the inner surface of the skull (namely, the brain) shrank faster than the outer one, leaving a space in between filled by a 'spongy substance'.[15] Although they estimated this divergence at one-eighth of an inch, phrenologists were reminded by experience that the analysis of younger heads (particularly those turned by flattery) tended to be more successful. Otherwise, there was little conjecture involved: a man's character was almost as obvious as the shape of the bumps on his head. Good phrenologists never automatically related intellectual ability to the total size of the brain, for it was a mistake to expect one person to be equal in intellect to another 'merely because their brains were equal in sheer magnitude'.[16] In Regency England it was still a commonplace that large heads meant large wits, but phrenologically this was not always the case. What really mattered was the relative size of the various organs to each other.[17] Broad and noble foreheads were indicative of intellectual prowess only because the 'Reflecting faculties' (*comparison, individuality, causality* etc.) were located there and compared favourably with the rest of the head. An accurate analysis was necessarily comprehensive, with measurements of all the faculties. Ordinarily phrenologists preferred to make measurements with calipers, after which the faculties were further described as 'very small', 'moderate', 'full' or 'very large'. If the subject was unwilling or had no time to submit his head to the calipers, the phrenologist had to content himself with making a quick visual analysis, perhaps giving special attention to the 'Reflecting Faculties' of the forehead.

Whether they used their measuring tools or estimated a man's moral powers from a distance or even from a portrait, Spurzheim and his British friends were accused of trying to judge fruit by examining the tree itself. 'We do not look at the tree and tell by its size how many and how good its apples will be', complained Thomas Halliwell in the *Examination and Refutation of Phrenology*.[18] Other anti-phrenologists echoed Halliwell's complaint by quoting classical proverbs, such as the saying (attributed to Livy) that 'There is no trusting the Features'. Even so, phrenology's answers were alluring because it was always possible to look for particular characteristics. If the public did not remember where all the faculties were located, they had none the less a certain fascination for some of the more unusual propensities. The horror of an ample *distructiveness* was clearly present in murderers with their 'contracted and threatening eye-brows and piercing eyes'; *acquisitiveness* in thieves was demonstrated by their 'longing, dissatisfied air and by their hands and heads advancing, ready to receive or lay hold of anything'.[19] Another infamous faculty was *amativeness*, which one anonymous pamphleteer (claiming to represent 'orthodox phrenology'), defined as the 'faculty of sexual attachment'.[20] There was little doubt about the role of this faculty. 'When it is small', the anonymous writer continued, 'the person will be cold and reserved; when large, fervently devoted to the opposite sex'. Like all the other mental faculties, *amativeness* fell victim to abuse, for if it were not 'controlled by decency and politeness [the expected effects of other faculties], this feeling may lead to libertinism and licentiousness'.[21] *Amativeness* was also one of the easiest faculties to identify. When 'large' it noticeably contributed to the thickness of the back of the brain nearest the neck, a point well illustrated by various writers (including Gall) who referred their readers to portraits of Henry VIII.

Phrenology thus presented a view of the brain which was the aggregate of measurable parts. The components did not all develop at the same time nor to the same degree, and here was the prime reason for differences among men. Mental attitudes were merely the result of brain structure which, although simple in itself, became rather complicated whenever several organs were operative at once.[22] Because

Gall and Spurzheim both wanted a universal science, phrenology's competence was remarkably wide. All creatures from the ostrich to nordic man were possessed of mental organs whose force and prominence the calipers could determine, although many of the organs found in human brains did not exist in animal ones. Whenever the *Phrenological Journal* carried articles on the *Philoprogenitiveness* of birds and fishes, it never failed to remark that the same general principles applied to man as well.[23] Phrenology helped to explain the process of mental derangement in all creatures and it was a defence of self-restraint. It was useful in choosing a spouse and it was a guide to the education of the young. But for all its diverse applications, phrenology as a physiology of the brain 'was not, and could not become, an exact science, and must ever remain an *estimative* science'.[24] Its enthusiasts were forced to this conclusion as soon as their *self-esteem* suffered by the delivery of too many faulty analyses. They therefore set practical limits, as did Spurzheim, to the application of their knowledge, and they did not pretend to tell whether a man spoke German, wrote poems, or committed murder. What they did undertake to do, according to the *Phrenological Journal*, was to estimate how far a man possessed the natural and inherent powers which 'under the *proper circumstances*, would enable him . . . to attain either to high or to moderate excellence in any . . . pursuit.'[25] It was not a man's history, circumstances and education which were stamped into the shape of his skull: only his natural tendencies. Phrenology's chief purpose was the discovery of the 'primary elements of intellectual and moral character, . . . and nothing else'.[26]

The process of discovery raised serious problems for the conscientious phrenologist. Was he not, for instance, running the risk of making *a priori* judgments of character? The problem bothered J. L. Levison, the leading phrenologist in Hull, who described a coach ride he once made in Lincolnshire. As usual he was interested in the mental powers of his companions, whose heads he scrutinised (apparently without resorting to calipers). Seated next to him was a 'respectable-looking individual', whose high forehead gave him a 'grecian-like appearance'. Levison remarked that his analysis of the man was essentially correct, for their subsequent conversation was animated and pleasant, which confirmed his subject's powers of reflection.[27] Levison discounted the possibility that he may have connected the man's 'respectable appearance' with his mental ability, although he admitted that this hazard often faced younger, less experienced phrenologists. Others were more honest. William Scott, a member of the Phrenological Society in Perth, wrote that he was as convinced of the truth of phrenology as any one could be, but confided that he was puzzled by cases in which phrenological analysis pronounced able bank-clerks to be 'natural hewers of wood and drawers of water' and vice-versa. Together with his friends Scott felt compelled to protest against such analysis as a 'species of tyranny'.[28] George Combe was generally ready to concede 'indecision or discrepancy' in analysis, although he claimed that the possibility of error only put phrenologists in good company.[29] After all, did not chemists often fail in experiments, and were not mathematicians known to err with their sums?

It was not the most convincing argument, but in any case mistakes were easily obscured. A certain amount of fast talking usually rescued the phrenologist from an awkward situation. If he declared that a man had an ample faculty of *benevolence* and it was then objected that the man had never done a kindly deed in his life, there was no cause for laughter and derision. For if the phrenologist knew what he was about, he might find extenuating conditions in the skull. Perhaps the large *benevolence* could not function properly because it was countered by equally powerful faculties of *combativeness* and *self-esteem*. Since there were more than thirty mental organs, phrenologists had plenty of room for manoeuvre; 'they must eventually hit upon some feature [of truth]', commented Halliwell, 'they cannot help it.'[30] This criticism, a favourite one among anti-phrenologists, held fast. For one thing, it implied that the phrenologist was merely an observant fellow who, whether he realised it or not, based his analysis on the costume and conversation rather than on the head of the subject. Anyone could do the work of a phrenologist, for corduroy trousers spoke volumes for character and so did a fustian jacket. Moreover, Halliwell's objection to the number of mental faculties raised one of the most fundamental issues which phrenologists had to face. Granted that the faculties explained the whole range of human emotions and attitudes, how could anyone be certain there were only twenty-seven? Or thirty-three? Why not two hundred and thirty-three?

> 'If phrenologists trace to VENERATION a love of antiquity and a tendency to approve of all that is old ... should they not give an organ for venerating what is new? And if there is an organ for loving children, why not [have one] for loving parents?'[31]

Knowing what he did of the history of phrenology, it was inevitable that Halliwell should have posed such questions. Spurzheim had already added to the twenty-seven faculties which Gall had located, while the more generous American phrenologists were to find many more again; these discoveries led anti-phrenologists to comment that the brain was obviously developing much faster than Gall had first expected. In this instance the standard defence was that phrenologists did not decide the number of faculties: 'they admit neither fewer nor greater than they find in Nature'.[32] It was, of course, always possible to reinterpret Nature.

Sensitive to criticism and not wishing their science to become a closed book, the phrenologists were flexible. Like Gall, they left many of the present issues and embarrassments to the resolution of future research. They fairly expected to find new faculties, in so far as several cerebral functions seemed mysterious. Following Spurzheim's example, phrenologists drew their cranial maps with some care, showing vague or alternate lines interspersed with question marks and encompassing regions of mental *terra incognita*. In the long run, however, this pleasant imprecision created new problems. How were the faculties, illustrated on fold-out maps and engravings, to be numbered? It was relatively easy to classify the organs; phrenologists agreed which were 'reflective', which were 'superior' and which were common to man and beast. But the names and numbers (digits often being used for the sake of convenience) differed from one school of phrenology to another. Gall's numbering system was not Spurzheim's; categories and subdivisions changed places, 'orders' and 'genuses' overlapped, and before long, nomenclature was the cause of schism among phrenologists. Nor was the exact location of the faculties settled beyond question. Herbert Spencer and his phrenologist friends disputed the exact whereabouts of *amativeness* and *wonder*; they even thought that the functions of the latter were 'too confused and indefinite' to warrant it a place among the faculties.[33] Questions about the number and location of faculties were never really settled, and as late as 1847 Combe's correspondence reveals a continuing battle over the location of mental organs. To foes of phrenology like the *Edinburgh Review*, these issues proved that the new philosophy was 'despicable trumpery' and 'thorough quackery from beginning to end'.[34] And yet the most important problem for phrenologists and their opponents involved not cranial cartography, but the determinism of phrenology. Was phrenology to accept man as he was? Was there no way to cure mental organs subject to perversion or simple disuse?

The problem is illustrated by a letter which Combe received early in his career from a man claiming to enjoy a good reputation as a portrait painter. The artist complained that his faculties of *locality* and *form* were rather small and that this was doubtless the reason for much of his inferior work. In the same way that a patient might ask his doctor whether an operation is really necessary, the artist asked Combe what could be done to remedy the deficiency of the two faculties.[35] The phrenologists never found it very easy to answer such questions, and the questions were put all the time. On the one hand, phrenologists emphasised the liberty of man, in Spurzheim's words, 'to will for or against anything'. In this fashion human determination might surmount the natural limitations of the human mind. On the other hand, man could not reverse his deficiency entirely; he could not develop a 'very large' faculty where only a 'very small' one existed before; nor was he able to reduce an animal propensity (such as *amativeness*) of 'very large' dimensions to one which was barely visible. Man had to be satisfied with only a partial reduction or growth: he might develop from a 'very small' to a 'rather small' condition, but never to 'rather large' or 'full'.

Unfortunately we do not know whether the artist who wrote to Combe was satisfied with this explanation. He may have been more confused than before. How was one to make such fine distinctions, to decide what was 'large' or 'very large'? Here, it seemed, was one of those handicaps which Gall attached to the 'speculative' philosophy of Spurzheim and his friends. Whatever the flexibility of its jargon and description, phrenology could not escape the element of determinism: no single faculty – and certainly no brain – was liable to a complete transformation. To calculate otherwise was to fly in the face of Nature. The *Phrenological Journal* clarified official thinking on the subject by saying that

> 'We cannot essentially *change the character* of any natural feeling, so as to convert ACQUISITIVENESS into BENEVOLENCE, and our efforts are limited to *restraining* the different faculties from *improper* manifestations, and to *directing* them to legitimate and *beneficial* indulgence.'[36]

For the artist distressed by his powers of *locality* and *form*, phrenology offered only faint solutions. Man could improve only within the bounds of the character which Nature gave him at birth. Hope and determinism, in unequal measures, were the dual components of phrenology, and equivocation over the mental faculties was its most obvious weakness. But the main assumptions of phrenology nonetheless remained imposing: the belief in the brain as the agent of all ideas, composed of distinct, operative parts and knowable as any other part of the body – these assumptions were not easily refuted. Equally imposing was the comprehensiveness of phrenology. However unfinished its advocates admitted it to be, phrenology was the only systematic explanation of the human mind: the proper interpretation of its findings promised long-awaited answers about man's existence. In this light, the quarrels of the phrenologists over the exact number and location of the mental faculties did not mean very much. The fundamental principles of phrenology were absolute and unvariable, and they invested the science with a philosophic monopoly which (in the opinion of one of Combe's correspondents) bordered on tyranny. The *Phrenological Journal* put it more bluntly: 'If phrenology, be true', the *Journal* asserted in its introductory statement, 'all other systems of the philosophy of mind are false.'[37] Obviously the revelations of Nature and scientific anatomy did not call for modest conclusions.

With its precision and confidence, its skulls and casts, its answers and its tyranny, phrenology gained ground in Britain. The story of its progress after more than twenty years was finally told in 1836 by the famous English botanist, H. C. Watson. His account, the *Statistics of Phrenology*, was, as the title suggests, essentially a narrative of social arithmetic; it described the fortunes of the new science in various localities around the country and the wide opposition it often received. In 1836 Watson was one of many skillful defenders of phrenology in the British Isles, whereas in 1816 there had been a solitary champion, a man whom Coleridge found 'dense and likeable, but the most ignorant German I ever knew'.[38] The differences between Watson and Spurzheim were immense, although their enthusiasm was approximately the same. It was the German, however, who almost succeeded in spite of himself. Spurzheim was not an attractive figure intellectually; his personality and his command of the English language were both rather rough. In fact, he had hardly learned English when, in the months following the Peace of Paris, he hurried from one lecture hall to another as the first true missionary of phrenology in Britain. He was determined to transcend his disadvantages as a foreigner and he won, even from his critics, a certain respect for his drive and sincerity. Certainly, as far as the British phrenologists were concerned, he was a model for his boldness: he preached the new science as if he were fighting a crusade. Presenting his case before the Royal Colleges of Physicians and the Royal Societies of London, Dublin and Edinburgh, Spurzheim confronted scientists and philosophers on their own ground. He did not fear these men, nor did he retreat in shame before the tiny audiences which first heard him in Derby and Dublin. It was not long before he brought the creed of phrenology into the vernacular: while continuing his ambitious programme of lectures, Spurzheim issued an English edition of the *Système Nerveux*. For all this, his labour was only partly successful. Medical men politely received his remarks on brain structure, noting that he had been a colleague of Dr Gall, but many were dubious about the 'necessary connections' which phrenology made between 'cerebral departments and human actions'.[39] A few parsons were sufficiently convinced to bless phrenology from their pulpits, although many more denounced it. Actually the small number of adherents never worried Spurzheim; he knew that conversions would take time. Long before he departed for the American tour from which he never returned, Spurzheim was content to generate in his British friends a fervent audacity in meeting opponents while he himself presided over the fortunes of British phrenology.

In some respects Spurzheim's hard work resembled a continuous election campaign. His conscious intention was to bring his message to all parts of the country and to create for himself the reputation of a national celebrity. His hope was that London society would eventually accord him the sort of reception which Paris had earlier given to Dr Gall, and in this Spurzheim was to be disappointed. Perhaps his chief problem was that he was carried away by his own enthusiasm. So genuine and professional was his commitment to phrenology that he imagined the science was about to sweep the western world like a new religion. The encouraging signs were everywhere in the 1820s. In France the Phrenological Society of Paris was strong enough to offer a prize of 1000 francs for the best essay on the 'application of phrenology to metaphysical analysis'. In Copenhagen, the leading Danish phrenologist, Dr Charles Otto, was favoured by the king, who had attended Gall's lectures some years before. In the United States, the Phrenological Society of Philadelphia was established in 1822 (one year before the founding of a society in London), and the Phrenological Society of Washington included among its members the Surgeon General and the Attorney General; John Quincy Adams, J. C. Calhoun and Henry Clay, together with an array of congressmen, expressed interest in the science and agreed to the measurement of their heads.[40] Like other trans-Atlantic phenomena of the nineteenth century, phrenology involved an interchange of ideas and personnel. Early editions of Spurzheim and Combe appeared in New York and, when they could not visit Britain, the American phrenologists directed waves of correspondence towards London and Edinburgh. Not unreasonably, Spurzheim always saw himself at the apex of this activity, and it is not surprising that he should have remained a missionary until the end.

It was during a lecture tour of the United States in 1832 that Spurzheim died. His burial in Boston allowed his American followers to become the guardians of one of phrenology's most hallowed shrines. His powers of persuasion and his boldness had continued to impress people; his funeral committee included a respectful Josiah Quincy, president of Harvard, and the Boston Medical Association 'attended in a body'.[41] Spurzheim's efforts could not go unfinished, and even his death seemed to give phrenology a curious impetus. So it was natural that George Combe, now the leading light among British phrenologists, should proceed to the New World; he wrote to Mrs Maconochie in Van Diemen's Land that it was 'a sense of duty which calls me to do good at once for the people of the United States'.[42] Obviously Combe thought no less of his own stature as a philosopher than Spurzheim had of his, and it is probably unwise to consider the impact of phrenology among the British without first sketching the role of George Combe as the most notable publicist of the science anywhere.

For fifteen years before his arrival in New York, Combe had received letters from Americans who insisted that phrenological societies were springing up from Boston to the Carolinas. Phrenology, Combe thought, must be exceedingly popular by now; he prepared his wife for what he was certain to be a splendid welcome. Combe found only Nahum Capen, his Boston publisher, waiting for him when his ship landed. Eventually, of course, American phrenologists called on Combe to render due homage and respect, for they had no recognised leader of their own. There was only Charles Caldwell (1772–1853), a pupil of Benjamin Rush, who once described himself as 'a long-time advocate of phrenology: perhaps the only one of note in the United States'.[43] But Caldwell was somewhat isolated at his home in Kentucky, and George Combe was thus left to become phrenology's chief in the Anglo-Saxon world. The Americans clearly recognised him as such. To Combe came requests for lectures and promises of honorariums which would certainly have excited Dr Caldwell's *self-esteem*. In New York Combe contracted to do a series of six lectures for $50 (£10), and in other cities the rewards were even more enticing. In Boston, Horace Mann arranged for Combe to deliver three lectures; Combe insisted that all school teachers be admitted without charge. He could well afford this indulgence, for the series still brought him about $200,[44] and he was so gratified by his reception that he announced that New England contained some of the most excellent heads he had ever seen. With two of the region's most excellent public servants, Horace Mann and Governor Everett of Massachusetts, as Combe's hosts, New England must have seemed very promising ground indeed. In the capital, Combe was brought to the White House and introduced to President Van Buren, whose bald head revealed large faculties of *benevolence, self-esteem* and *love of approbation*. Combe was well satisfied with the results of his American mission and he had reason to be. He managed to cover a great deal of territory and his lectures were generally well-attended, well-published and well-subscribed. But it was not Combe's American trip, or his frequent travels in Britain and Germany, which brought him a vast repute. He was not extraordinary as a public speaker; he admitted that he was quite uninspiring at the podium. As Combe's correspondents were already aware, his real force, his spirited logic and his flair in argument were all far more abundant in his writing, and it was his published works which truly won for him the full measure of a world-wide fame.

Combe's first work, the *Essays on Phrenology*, appeared in 1819. It was a timely and straightforward defence of a science which had already caused a great deal of argument — a situation suggested by the rest of the title, *An Inquiry into the Principles and Utility of the System of Gall and Spurzheim, and into the Objections made against it*. Even though the book did not prove the last word which Combe hoped it would be, several later editions were issued under the shorter and more convenient title, *A System of Phrenology*. In the years that followed, Combe was to produce many more volumes, and a few of them said nothing about skulls and faculties. The remarks which the *Edinburgh Review* made in 1826 of the *System of Phrenology* might also be applied to the rest of Combe's works. The *Review* found the *System* 'long, sober and argumentative'. Such were the virtues of writing in the 1820s, and Combe was read because he assumed his audience knew nothing of a subject: he proceeded step by step to develop every point lest it ever be doubted. Because he dealt with all topics in this way, Combe frequently enjoyed good reviews. He was complimented for his writings on the currency question, for example, because he explained a difficult subject in terms which the public readily understood; *The Times* thought his pamphlet on the *Currency Question* 'a real service to the commercial public ... fulfilling everything that could be required'.[45]

Combe's style or writing (whether in his published works or in his private correspondence) also tended to be moralising and didactic. But these, too, were acceptable qualities. The reading public of the 1820s and 1830s was in a decidedly evangelical temper; they were equally accustomed to long and quarrelsome exchanges in the press and they were intrigued to read of 'refutations refuted'. The increasing volume of documented reports from parliament and the statistical societies had produced an argumentative frame of mind, and George Combe was no exception. His defence of phrenology usually resembled the format of a problem in geometry. He began with the 'given principles' of anatomical and medical evidence derived from dissections; he then cited well-known examples of various mental characteristics. The reader was finally invited to see for himself how 'noble-looking heads' were usually indicative of great intellects. It was a neat inductive case, depending upon the unchallenged first principles, the researches of Gall and Spurzheim. As a philosophical construct, phrenology may have been a house built on weak foundations, but Combe's arguments were stylish enough. And there was no denying that style counted for much; it was, as the anti-phrenologists admitted, sufficient to convert many readers, or at least dispose them favourably, to phrenology.

> 'It is impossible,' confessed the Edinburgh *Review*, 'not to admire the dexterity with which he [Combe] has evaded the weak, and improved the plausible parts of his argument, and the skill and perseverance he has employed in working up his scanty materials. into a semblance of strength and consistency.'[46]

The irony of such praise was not meant to please phrenologists, but there was the admission that if they presented their science as clearly and as logically as they could, a certain measure of success was almost guaranteed. Hewett Watson had no doubt of this. As he later noted in his *Statistics*, the mere fact that 1500 copies of the *Elements of Phrenology* were sold in ten months was 'convincing proof' that both Combe's style and message were getting through.[47]

The books which George Combe wrote on phrenology almost all date from the 1820s and 1830s. In the 1840s he turned his attention more particularly to issues of education, religion and prison reform; in the 1850s, to the currency question, biography, and problems of world peace and empire. Throughout this span of time and topics, Combe's style underwent surprisingly few mutations. His final drafts depended more on long-formed impressions and upon common sense than they did upon months of research and criticism. Possibly Combe believed that his own mental faculties afforded him special insight and sound judgment which Nature denied to others; in any case, he approached each of his works with a vigourous sense of his own authority. His attitude is discernible in the treatment of currency and educational reform, but more obvious still in the works on phrenology.

Combe was convinced that all attempts to supersede phrenology, or to dilute it by advocating groups of faculties rather than individual ones, were wrong and destined to failure. He professed a dislike for Catholic and Calvinist narrow-mindedness, but he would hear of no challenge to orthodox phrenology. Phrenology was God's truth; the faculties were His; the divisions and categories were His; and 'no human sagacity can improve upon His plans.'[48] For his dogmatism Combe was not beyond reproach, even from his friends. Spurzheim, who believed that the *Essays on Phrenology* was 'the most able defence of phrenology in the British Empire, certain to make a sensation among those who like truth',[49] rebuked Combe for speaking 'so decidedly' of various mental faculties. 'Phrenologists', Spurzheim warned gently, 'cannot be too much united in doctrines and assertions. Let it be said, as it is, "Dr Gall thinks this", and "Dr Spurzheim thinks that", and let experience decide who is right.'[50] However, it was precisely the unity of phrenology which Combe desired to uphold in all of his publications. He once referred to himself as the protector of orthodoxy, the guardian of the only true theory of man, of the Great Answer which Aristotle, Bacon, Locke and Dugald Stewart toiled to attain but without success.[51]

As custodian of phrenological truth, Combe increasingly relied on the power of the pen. He retired fairly early from his law career in order to have more time for writing. He wrote much of the *Phrenological Journal* himself, although officially he was only one of several editors. He addressed 'open letters' to provincial newspapers, responding to the opponents of phrenology. Some of these letters were reprinted as tracts or small pamphlets but few survive. Those extant, such as the 'Letter to Francis Jeffrey' (a determined anti-phrenologist), or the 'Correspondence of 1828', are invariably partisan morsels, snappish exchanges on the merits of the new science. Many who read Combe's full-scale works such as *Moral Philosophy* (1840), *Essays on Phrenology* (1819) or above all the *Constitution of Man* (1828) felt that Combe was putting a damper on the development of the new philosophy and also taking a great deal of credit to himself. George Jacob Holyoake, for one, declared that while Combe accomplished for phrenology what Paley had for theology, both men 'would have stood higher in the estimation of their readers had they owned what they owed to their predecessors'.[52] This criticism of Combe was not very fair. He was, it is true, proud of the reputation he eventually gained and somewhat complacent because of it, but he was not a plagiarist and he never hesitated to acknowledge his debt to Gall and Spurzheim.[53] At any rate, it did not matter to the reading public whether or not Combe was entirely original; like Holyoake, they believed Combe was phrenology's 'greatest expositor' and for this reason they were ready to bestow upon him either honour or abuse.

It has already been mentioned that several of Combe's works bore no immediate testimony or connection to phrenology. The currency essays clearly fall under this heading, as do many of his essays on secular education and recollections of foreign travel. The non-phrenological works posed a special problem, for while Combe desired respectable sales he faced immense competition. That he should have succeeded was due to the fact that his writings (phrenological or not) were designed for the widest possible audience. This much is clear in his preparation of the *Notes on America* (1841), a handsomely bound work in two volumes. Combe did not expect it to sell well among the upper classes because (as he explained to Cobden) 'my reputation does not stand very high among the rich'. He therefore considered producing a popular edition, 'printed in double columns for the multitude, if it were thought likely to be of any use to them.'[54] He need not have worried, for the *Notes on America* enjoyed a good circulation. Lord John Russell believed Combe's remarks were those of 'a candid and honest observer';[55] Combe's style was such that, as an editor of the *Economist* later confided, 'it is no small honour to be favoured with the writings of Mr Combe'.[56]

Thanks to their stock of common sense and clear expression, Combe's non-phrenological volumes sold tolerably well. But none of them ever captured the vast public which belonged to his great phrenological treatise of 1828. The title of this volume, the *Constitution of Man considered in relation to external Objects*, was not then as unwieldy as it may seem today; more positively, it represented the last word on phrenology as a philosophy of man. It was, in fact, the basis of orthodox phrenology, the fullest expression of that science begun by Gall and Spurzheim and denounced as foreign quackery. What accounted for the immense attraction of this work? How could a thousand copies be sold in a single week, and how was it possible for Harriet Martineau to rank it among those ubiquitous books, *Pilgrim's Progress, Robinson Crusoe* and the *Bible*?[57] Part of the answer lies in the talents of the author, his devotion to phrenology and his aggressive style; the bulk of the answer lies in the advent of phrenology itself as a popular science and philosophy. We must now explore the reasons for its wide appeal. We must, in other words, see what the *Constitution of Man* reveals of the intellectual constitution of the society who read it.

Notes

1. Edinburgh *Review*, XXV, no.49 (June 1815), p.228.
2. F. J. Gall, *On the Functions of the Brain and of Each of its Parts*, trans. Winslow Lewis, Jr. (Boston, 1835), I, p.57f.
3. *Ibid.*, p.55.
4. *Ibid.*
5. R. M. Young, *Mind, Brain and Adaptation in the Nineteenth Century* (Oxford, 1970), p.24.
6. Gall, *op. cit.*, III, p.82.
7. Erwin Ackerknecht and Henri Vallois, *Franz Josef Gall, Inventor of Phrenology and his Collection*, trans. C. St. Leon (Madison, Wisc., 1956), p.9.
8. In a letter to George Combe, dated 13 March 1821, Spurzheim gave a brief sketch of his education and career and his approach to phrenology. NLS 7207/75–76.
9. George Combe, *Essays on Phrenology; or an Inquiry into the Principles and Utility of the System of Drs Gall and Spurzheim, and into the Objections made against it* (Edinburgh, 1819), pp.140–55.
10. *Ibid.*, p.201.
11. George Combe, *Elements of Phrenology* (4th ed.: Edinburgh, 1836), pp.59–60.
12. *Ibid.*, p.79.
13. *Ibid.*, p.161.
14. Combe, *The Life and Correspondence of Andrew Combe*, p.416.
15. Combe, *Elements*, p.12.
16. *Ibid.*, p.151.
17. *Ibid.*
18. Thomas Halliwell, *Examination and Refutation of Phrenology* (Dunedin, 1864), p.12.
19. [anon.] *Phrenology, and the moral Influence of Phrenology* (London [1835]), p.8.
20. [anon.] *Orthodox Phrenology* (London [n.d.]), p.5.
21. *Ibid.*
22. Combe, *Elements*, p.7.
23. *Journal*, II, no.5 (1824), pp.20–22.
24. Combe, *Relation between Science and Religion*, xvi.
25. *Journal*, I, no.4 (1824), pp.561–2.
26. *Ibid.*, p.559.
27. *Ibid.*, VI, no.21 (1829), p.135.
28. Scott to Combe, 16 August 1826. NLS 7218/77.
29. Combe, *Elements*, p.153.

30. Haliwell, *Examination and Refutation*, p.17.
31. *Ibid.*, p.31.
32. Combe, *Elements*, p.177.
33. Herbert Spencer, *Autobiography* (2 vols., London, 1904), I, p.246.
34. Edinburgh *Review*, *op. cit.*, p.227.
35. Unsigned letter to Combe, 23 December 1822. NLS 7208/52.
36. *Journal*, I, no.2 (1824), p.222, referring to 'Mr Owen's New View of Society'.
37. *Ibid.*, no.1 (1823), viii.
38. *Table Talk and Omniana of Coleridge*, 29 July 1830.
39. The *Dublin Journal of Medical Science*, XIII, no.38 (1838), p.179.
40. The *Phrenological Journal* (XIII, no.65, p.314f) once described the cerebral traits of various American politicians.
41. R. E. Riegel, 'The Introduction of Phrenology to the United States', *American Historical Review*, XXXIX, no.1 (1933), p.75.
42. Combe to Mrs Maconochie, 20 August 1838. NLS 7388/77.
43. Caldwell to Combe, 25 June 1821. NLS 7206/48-49.
44. George to Andrew Combe, 27 September 1838. NLS 7378/78.
45. *The Times*, 4 March 1856, p.11.
46. Edinburgh *Review*, XLIV, no.88 (1826), p.253. The *Review's* article on the *System of Phrenology* ran over sixty pages.
47. Hewett C. Watson, *Statistics of Phrenology: being a Sketch of the Progress and present State of that Science in the British Islands* (London, 1836), p.15.
48. Combe to Margaret London, 24 December 1845. NLS 7390/244.
49. Spurzheim to Combe, 22 October 1819. NLS 7204/158.
50. Spurzheim to Combe, 16 November 1824. NLS 7214/63.
51. George to Andrew Combe, 4 November 1822. NLS Draft letters.
52. George Jacob Holyoake, *Sixty Years of an Agitator's Life* (2 vols., London, 1892), I, p.65.
53. Combe, *Constitution of Man*, vii.
54. Combe to Cobden, 3 March 1841. BM43,660/7 (C.P. XIV).
55. Russell to Sir James Clark, 24 February 1842. NLS 7266/17.
56. A. Webster to Combe, 9 September 1847. NLS 7288/144.
57. Martineau, *Biographical Sketches*, p.265.

Chapter III A Respectable Concern

British phrenologists were never under any illusions about the difficulties they faced in persuading people. Spurzheim did not have an easy time in Britain himself, and he could never quite understand why there seemed to be more antagonism to his science in that country than in France or Germany. The first fifteen or twenty years after he began his work were particularly rough. Enemies arose on all sides. Respected surgeons, members of the Royal Academies, churchmen and journalists: all were eager to engage phrenologists in public debate. Sometimes the discussions were amusing; often they were fierce. There was much personal abuse, especially in the press. But the amount of knowledge then available saw to it that the arguments on both sides were rather sketchy. Indeed, one can almost say they were equally flimsy. And what was the final verdict of those educated men? Not surprisingly, it was given by a clergyman, the Reverend Brewin Grant, a longtime foe of the science. When he took part in a debate in 1850, Grant enjoyed at least the advantage of historical perspective. The problem with phrenology, he said, was that it did not rise 'from the highest circles of science' but from men of the 'lower orders of the intellect'. He much preferred dealing with the 'aristocracy of the intellect' and, with a glance at his phrenologist opponent, he assured the audience that 'Dr Gall was not one of that aristocracy'.[1]

It was a technical point, perhaps, but the Reverend was right. By 1850 practically no-one remembered Gall or the work he had done. His reputation, considerable only forty years earlier, had not survived. This very fact vindicated what Grant called 'the cream of British literature' — the Edinburgh *Review*, the *British Quarterly* and other journals which opposed Gall and Spurzheim from the beginning. But the intellectual aristocrats were not always just. Occasionally they fell to the temptation of lampooning phrenology as 'continental quackery' and very often they rejected it simply for its non-British origins. Not even Charles Bell, whose research was to prove far more durable than Gall's, was entirely above prejudice. Giving a paper before the Royal Society of London in 1823, Bell attacked phrenology as 'foreign' and said that the work in the medical schools 'of this kingdom should be distinguished from those of other countries'.[2]

Bell's words carried much weight, but not very far. His contempt of phrenology had little or no effect upon those educated persons who still suspected that a man's head or his appearance pronounced judgment upon him. Nor did Bell dissuade those who, while granting that phrenology wanted more proof, believed that the science involved more than mere head reading. In a general way, all through the nineteenth century, phrenology appealed to savants and literati everywhere. Balzac and George Eliot looked to the contours of the head for hints of character; so too did Baudelaire and Marx. Phrenology (like physiognomy) helped the believer to understand the relationship between physical and moral structure – between, as it were, a noble countenance and a clear moral conscience. Phrenology lent its advice to anyone in a position of responsibility. To the medical practitioner, it drew attention to the condition of the brain and the moods of the patient. To the employer, it said more than any character reference or testimonial. To the parent, phrenology was a guide in the social and moral training of the young. A science which was able to accomplish all these things, wrote Robert Macnish in the *Introduction to Phrenology*, cannot be a trivial one, for it exposed the grievous errors of contemporary justice, education, and habits of living, while it also suggested the remedies.[3]

There are probably two discernable aspects to the popularity of phrenology among educated people in early Victorian Britain. Both aspects concern phrenology more as a philosophy than as a science, but this is hardly surprising when we realise what Spurzheim and the Combes did with the craniology of Dr Gall. The first aspect involves the intellectual methods and mechanics of phrenology; the second involves its remarkable harmony with existing philosophical attitudes. We shall, therefore, encounter phrenology not so much as a passing fad but as a scientific philosophy applied to the whole progress of man: a philosophy which Harriet Martineau (quite undaunted by the 'intellectual aristocrats' of her day) once called 'a boon to the multitudes, high and low'.[4] The enthusiasm of the 'low' must be left to another chapter. Here we are interested in the following which phrenology gained in polite society – and particularly among the men, physicians, surgeons and scientists, who might have rubbed shoulders with Charles Bell when refreshments were served after his paper to the Royal Society.

1

After reading the *Constitution of Man*, the young secularist George Jacob Holyoake praised Combe's essay as the 'new gospel of practical ethics'. The substance of the book was practical and so was the approach: a successful approach, in Holyoake's estimation, because it was the 'proper blend' of modern science and traditional morality.[5] For many thoughtful people, phrenology's advantage lay in this blending, for the science was comprehensive and at the same time uncluttered. True, there was no end to the illustrations which phrenologists could give about different facets of character; but the examples were fairly straightforward and usually based on studies of individual conduct recorded by reputable persons, such as physicians, parochial officers or biographers. These people provided the facts (as historians were supposed to do for Henry VIII or as wardens of hospitals and asylums were supposed to do for their inmates); phrenologists were left to interpret the facts scientifically. 'Case studies' were therefore very important to them. Early in his career as a phrenologist Combe pointed out that character analysis was not achieved by anatomical dissection alone:

> 'The anatomist [Combe wrote in 1819] might dissect the olfactory nerves, till his eyes grow dim with age, and never discover that these nerves perform the functions of smell ... He might dissect every organ of the body, but could never, from the anatomical structure of any one of them, infer the functions it performs in the living body.'[6]

The same principle applied to the departments of the brain, where the most accurate judgment, confirmed with calipers, derived from a reading knowledge of case studies and personal records.

To the modern reader many of the studies collected by phrenologists have the quality of the 'human interest story'. Actually the facts of the cases were usually genuine enough – it was their interpretation which brought criticism. There was a certain beauty in the quest for 'simple facts', and simplicity in philosophy was the order of the day. Plain examples were wanted, not the endless ramblings of 'feel-osophers'. So great was the premium on an intelligible modern philosophy that phrenologists and anti-phrenologists constantly tried to label each other as old-fashioned for relying too heavily on the philosophical methods of the past. In his long article in the *Encyclopedia Britannica*, for example, Dr Peter Roget accused phrenologists of arguing 'from the slippery ground of analogy', although there was not much he could do about the authenticity of the records which Gall and Spurzheim had drawn upon.[7] The usual course was to dismiss the records (so often compiled by non-phrenologists) as a series of coincidences; but was that really safe? Many surgeons and scientists thought not; the case studies of 'observable phenomena' were not so easily rejected, whatever the vagaries of phrenological analysis. Indeed, medical men frequently praised phrenology for its emphasis on physical and neurological factors in human character and for attempting to use the information which traditionalists had always ignored.

Quite apart from its conclusions, phrenology was attractive for its treatment of a vast store of information. It was now fashionable, and had been for some time, to believe that knowledge was cohesive and systematic; Providence had ordained a place for every discovery of the human mind. The nature of human knowledge was that it came gradually but always lent itself to organisation. Nineteenth-century phrenologists were not in the least outlandish for their view that Truth equalled the sum of many departments or categories, all of which could be identified. It has been said that 'classification' was one of the intellectual penchants of the age: it led to Darwin's *Origins*, Spencer's *First Principles*, Mill's *Logic* and George Combe's *Elements of Phrenology*.[8] When Dr John Epps demanded that man be considered phrenologically, the implication was that an analysis could be made of the main groups or categories of mental faculties: it was possible to reach a grand average for the 'reflecting powers' in contrast to the 'animal propensities'. And for a truly complete picture of the human condition, the classification of mental powers was augmented by one of the physical. In the *Elements of Phrenology*, Combe accepted the age-old explanation of the 'Four Temperaments', and in their correspondence phrenologists often described individuals as sanguine, bilious, lymphatic, or nervous.[9] The methods of character analysis thus involved two separate (but not wholly independent) judgments: the first dealing with the physical signs of the brain itself, and the second with the facial features, general state of health, and (not least of all) the history, if available, of the individual's moods and dispositions. There was therefore a place or category for everything that could possibly be known about human nature; believing that mental traits could be classified just as physical traits had been, the phrenologists claimed to finish the portrait of man begun by the ancients.

The wisdom of phrenology had yet another apparent advantage. In addition to its clear organisation and its foundation in observable facts, it was plainly expressed. The devotees of phrenology considered it an accurate but simple philosophy, unencumbered by the difficult language of metaphysics. Much of phrenology's success, among men of station as well as the general public, derived from a weariness of metaphysics, a weariness already very common by the end of the eighteenth century. The philosopher Thomas Reid openly confessed, in his *Essays on the Intellectual Powers of Man*, that he was baffled by the invention of 'an improper phrase to express what every man knows how to express in plain English,'[10] and like Reid phrenologists did not hide their contempt. Combe referred to the 'barbarous jargon and logic of metaphysics',[11] while the American physician Charles Caldwell pronounced metaphysicians and alchemists equally worthy of scorn. Phrenologists considered metaphysics a particular obstacle to moral philosophy; they thought its imprecision gave rise to countless theories which 'have bewildered the most enlightened nations'.[12] There was ample proof of such confusion in Germany, which for most phrenologists was the most enlightened corner of Europe. There Hegel's 'occult science' (so described by leading phrenologists) represented the most current failure of metaphysics. It seemed to defy all human understanding — or at least it defied that of George Combe, who decided that 'peculiar talents and life-long study' were necessary to understand it.[13] The contrast with phrenology was obvious. Phrenology stood alone, independent of its founders and free of the intricacies of old metaphysics.

Or so the phrenologists thought. In reality they were embarrassed by what can only be called metaphysical problems. Some of Combe's correspondents complained of the imprecise distinctions made between emotions, instincts and the faculties themselves; disagreements were bound to arise in the study of an 'estimative science'. Still, phrenologists were justified in claiming to have a philosophy expressed in clear modern English. In this connection the very names given to the faculties are revealing. Several of the expressions (notably 'Benevolence' and 'Self-Esteem') frequently occur in various sorts of late eighteenth-century literature, and even among the opponents of phrenology there seems to have been very little confusion about the meaning of names of the faculties. But the familiar nomenclature and easy jargon was never enough; nor were the intellectual methods of phrenology, taken together, the most alluring feature of the new science. The substance of phrenology had a much more familiar ring to it.

The affinity of phrenology to the whole pattern of late eighteenth- and early nineteenth-century ideas is truly remarkable. The first chapter of the *Constitution of Man* established the style not only for that popular volume but for much of what phrenology had to say. The chapter, entitled 'On the Natural Laws', dealt with the rationalist concept of creation well known to the philosophers of the last century. It assumed laws which were eminently knowable and reliable. Man's recognition of these laws accounted for the great bulk of his social progress; his respect for the 'relationships which exist in Nature' insured him of good health, social happiness and an appreciation of the Divine Plan.

To speak, as the phrenologists did, of 'relationships', of organic and natural laws and of the 'all-powerful Cause' was to reveal the most definite philosophical commitment. Rationalism had not disappeared with the previous century; as so many of the published titles of the 1820s and 1830s suggest, there was still a wide enthusiasm for the 'general laws of Nature' and man's ability to know them. Indeed, it is quite possible that among educated people there was more talk about 'natural laws' than ever before. Rationalist ideas were presented in the most agreeable tones, whether in academic circles or in the established journals. The reverence was almost blind. No-one (including George Combe) bothered to enumerate the laws; if pressed to give an example, they almost invariably spoke of the 'laws of gravity' and nothing else. Rationalism had become an intellectual comforter – or perhaps it always was: first for the elite of the eighteenth-century *salons*, and now for the university-trained middle-class professions. At any rate, the social science of phrenology supposed that the laws of the social world were originally those of Nature and, more than that, that the mental faculties of man corresponded to the natural laws of the physical world.[14] Reason itself was only a comprehensive term for the 'reflecting faculties' and the 'higher sentiments'; the 'Voice of Reason', to which men of the Enlightenment constantly deferred, was simply the dutiful organ of *conscientiousness*, implanted by the 'Author of the Universe'.[15] To man's understanding of his world the phrenologists added a rational exposition of man himself, and they thought it their duty to point out that the human mind was no exception to the whole pattern of Creation.[16]

In addition to its underlying rationalism, phrenology bore a close resemblance to at least one recent school of thought. Combe and the early phrenologists were well acquainted with the Scottish school of philosophy which had developed around Reid and his pupil Dugald Stewart. The ideas of Thomas Reid (1710–96) were influenced by the study of physiology, and he even arrived at an interesting list of twenty-four 'active powers of the mind', some of them bearing the phrenologically acceptable names *duty, judgment, imagination*, and *self-esteem*. While the publication of Reid's essays certainly preceded any of Gall's work, there is no evidence that either Gall or Spurzheim was acquainted with the essays. It is more likely that Gall located and defined the twenty-seven faculties from his own observations, and that some phrenologists welcomed his research as the practical fulfilment of Reid's theory.

A more striking precedent was the work of one of Reid's disciples. The career of Henry Grey Macnab was in many respects a curtain raiser for that of George Combe. Born in 1761, Macnab began one profession (medicine) which he changed in favour of another (education). When peace was restored in 1815 Macnab resumed his travels and investigated the ideas of Robert Owen, which he incorporated into his educational schemes. In 1818, one year before the appearance of Combe's *Elements of Phrenology*, Macnab published his *Analysis and Analogy in Education*, in which he treated the frequency and the degree of various 'mental powers', concluding that knowledge was resident in 'powers which are possessed ... before men even acquired habits of reasoning'.[17] Such conclusions were more akin to Spurzheim than to Owen. Macnab's treatment of the 'General Laws of the Constitution of Man' may have suggested to Combe more than a clever title – had Combe read Macnab.[18] There is no evidence that he did. There is evidence, however, that the intelligent onlooker of 1820, having perhaps read Reid's *Essays* and known about Macnab, Spurzheim and Combe, came to the conclusion that craniology was not as alien or as outlandish as the Edinburgh *Review* said it was.

In fact, the phrenologists themselves hurried to clear up the confusion. As soon as they had founded their own journal, they rejected the notion (apparently already common) that they were an offshoot of Reid's school. The mental science of Reid and his followers was incomplete; it might have stumbled upon the names of mental powers, but it could 'never account for or explain the real movements or elements of any one mind, and far less the characteristic differences which exist between one mind and another'.[19] Reid and Macnab did not go far enough because they spoke only of functions of the mind, completely neglecting the physical facts of cerebral organisation. For phrenologists this omission was terrifying: if a modern philosophy made only tentative and brief use of physiology, it would eventually have to lean back upon the old metaphysical arguments and all the endless speculation of traditional logic. Phrenologists claimed to avoid that *impasse*. So forcefully did they stress the physical and discernable nature of mental powers that they sought a clean break with the abstractions of the past. In doing so, they were accused of materialism, just as Dr Gall had been; but that did not matter. Their consolation was that they were truly practical and utilitarian philosophers, while the school of Common Sense was not. In the essays of the *Horae Phrenologicae*, Dr Epps, who had long sought a practical philosophy, underlined this position in plain terms. Happiness, he declared, was man's real goal, and the operation of certain mental faculties produced it. Reid's 'powers' and 'functions' were almost worthless; human knowledge and human happiness came only from considering the human condition phrenologically.[20]

Avoiding metaphysics and unfamiliar language, phrenology was a tidy mental science for the gentleman. Admittedly there were still questions to answer, but these were left to time and to a methodology which embraced both traditional and revolutionary techniques. As phrenology was an unfolding system of facts about human character, there was no reason to be too dogmatic. Spurzheim had acknowledged this early, partly because he discovered a number of new faculties; he wanted phrenology to be a flexible, open system of moral philosophy. Dr Epps, wanting the same thing, emphasised that phrenologists should 'refuse the name of dogmatists'.[21] In view of phrenology's evidence (or lack of it) this was the most sensible attitude to take, and one which was not without its attractions to educated persons. A philosophy with scientific pretensions, speaking the layman's English and connected to the most digestible ideas of recent times: that in itself was alluring to those who read more than newspapers and novels. But a system of knowledge, based on hundreds of visible case studies and honest enough to admit of further development: *that* was almost compelling. Such were the apprent advantages of an open and mobile science. Men of all professions found these features to their liking: lawyers, educators, chemists and writers. Intellectually, however, phrenology's demand for attention was always felt more by the men of medicine and science.

2

When they published their long and vehement attack on phrenology in 1815, the writers of the Edinburgh *Review* were worried that Dr Spurzheim might be able to 'delude a few of the medical tribe, who are naturally looked up to as judges in questions of this sort'.[22] The anxiety was well founded. From the beginning phrenologists counted on the support of the medical profession, and they made a special effort to put their case before physicians and surgeons. It was fortunate for the new science that the Combe brothers and several other advocates of phrenology made their home in one of the most renowned centres for the study of medicine. By the end of the eighteenth century the fame of the Edinburgh medical school was world-wide. Perhaps only a third of its annual graduates were Scots; the rest came from all over Britain and Europe.[23] Many who were later expected to support phrenology did medical or anatomical studies at Edinburgh in their youth: among them were John Epps, Sir James Clark, John Conolly, Sir James Kay-Shuttleworth and George Birkbeck. The Combes did what they could to cultivate the friendship and goodwill of these and many other promising young medical men. What they hoped to achieve was the sort of commitment they won from Dr William Moffat, physician to the Belfast Hospital, who printed (at his own expense) handbills which he circulated among his colleagues in Ireland. The advertisements included a note on the medical value of phrenology and advised every doctor who wished to practice 'with pleasure and success' to be a phrenologist.[24]

The number of medical men to pledge themselves as boldly as Dr Moffat was never as great as the Combes hoped. Many physicians and surgeons could not decide. Some accepted phrenology simply because they thought that Gall and Spurzheim were quite sensible in the rest of their physiology. Besides, there was no established orthodoxy in dissecting brains; the organ could be cut up in any number of ways, much like the skin of an orange. The easiest position to take was one of complete scepticism, which was to be Dr Roget's position in the *Britannica* article of 1842. Roget said that the anatomy of the brain is so complex 'and so void of apparent adaptation to any purpose we can understand, that it will suit any physiological system nearly equally well'.[25] Of course not everyone escaped in this fashion. Even in the 1850s there were some medical men who thought Roget was being more defeatist than sceptical; they believed that phrenology was onto the truth. There was, therefore, a fair number of the profession who subscribed to phrenology from the beginning, and they comprised very sizeable proportions of the local phrenological societies. The society in Edinburgh, for example, numbered about 110 persons who regularly paid their annual dues of one guinea in the early 1830s, and of this number Watson's *Statistics* listed nineteen as 'medical men'.[26] Dr Roget might have wished it otherwise, but the fact is that his profession was seriously divided over the merits of phrenology.

This was certainly true of the medical journals. Several, including the Edinburgh *Medical and Surgical Review*, the Dublin *Medical Journal*, and the *Medical Gazette*, joined *The Times* and other papers in condemning the 'bumpologists'.[27] Several other professional journals, however, were much more polite. The *British and Foreign Medical Review* was usually friendly; the *Lancet* (already described in the late 1820s as the 'Koran of the medical profession') devoted much space and praise to the work of the London Phrenological Society and the books of George Combe. The *Medical Chirurgical Review* was still more generous: it openly declared its admiration for the way Spurzheim had 'outlived the illiberal treatment' which his enemies had heaped upon him for years. More important, in the *Review's* opinion, was Spurzheim's proof that the 'manifestations of the mind depend on the organisation of matter'.[28] For that point alone, Dr Spurzheim (and phrenologists generally) were entitled to a considerable amount of goodwill from the profession. Sometimes physicians would speak favourably of phrenology in their own books (as Dr Parry did in the *Elements of Pathology*); more often support was expressed by joining local phrenological societies or by subscribing to the papers which these societies printed.

The division was also evident in the public debates which phrenology generated. Open meetings at medical schools and in civic assembly halls, called to discuss phrenology, were common in the twenties and thirties. In Edinburgh Dr Andrew Combe participated in the debates which lasted several consecutive nights in November 1823, each debate drawing a greater audience. It was not always possible on such occasions for surgeons to escape the issue. And they often found that the arguments of clergymen and philosophers were not their own. The most notable example was the feud which raged between Dr Spurzheim, during one of his visits to Edinburgh, and Sir William Hamilton. The dispute was carried in the *Caledonian Mercury* in January 1828, and it was an event which surgeons talked about for years.

Hamilton was, of course, a decided foe of the new science, which he suspected of being irreligious. As a younger man he had for a time studied medicine and, when phrenology first appeared, he even dissected a number of brains to see for himself whether there was any truth to the theory. He decided there was not. But in 1828 neither the public nor the medical profession was willing to accept his word as final. Certainly the intellectual community esteemed Hamilton; but after pursuing law at Oxford and a bit of medicine and literary criticism at Edinburgh, some wondered if perhaps he had spread his talents too widely. By profession he was a philosopher rather than a surgeon; what was worse (and a little suspicious) was that he refused to attend the dissections performed by Spurzheim. The press berated Hamilton for not appearing to do battle, and the community generally felt that phrenology had won the day. The *Scotsman* was much impressed by Spurzheim whose character, 'full of cordial sincerity and philanthropy, commanded respect as well as attention'. More important, the *Scotsman* continued, was Spurzheim's 'obviously extensive knowledge of comparative anatomy and physiology; and from the manner in which he handled the brain, before a numerous *medical* and genteel audience, we are satisfied that his reputation as an anatomist and physiologist has not been carried too high'.[29]

While Spurzheim enchanted his audience in the Clyde Street Hall, Sir William made another tactical error. He declared in the press that he was quite sure the whole business of phrenology could easily be disproved 'by a single *practical* anatomist' — thus implying that some anatomists were able and others were not.[30] Ultimately one of the reasons why the two antagonists did not meet on the same stage was Hamilton's refusal to accept anyone as 'umpires' — again implying that he was a proper judge of surgical talent. The point which can be drawn from this argument in 1828 (and from Roget's criticisms in 1842) is that no anti-phrenologist could safely turn the question over to the surgeons. Judging by the articles in their journals and by their comments at public meetings it seems that for every surgeon who rejected phrenology there was one who condoned it. The special significance of the 1828 exchange was not its sour and inconclusive progress but the fact that before an audience 'largely medical and genteel' phrenology was not defeated. The issue was regarded as an open question and, in the absence of final proof on either side, each man was left to decide for himself.

Actually the outcome of such debates was not altogether vague. The intellectual advantages of phrenology were not lost on the medical profession. Here, after all, was a moral philosophy (perhaps the first in the history of man) which did some credit to the study of man himself and to the practice of dissection — a practice long disfavoured by the same clerical gentlemen who now, as a profession, opposed phrenology. Not until 1826 were degree candidates at Edinburgh obliged to study dissection formally, at a time when phrenologists were helping to change public opinion on the subject. Indeed the connection was frequently made between phrenology and anatomical investigation. Even the dissection of a criminal for purposes of anatomical study had become a matter of 'great inconvenience' in the academies for the number of outsiders it attracted. 'Thirty years ago none wished to look at the body of a murderer,' wrote Dr Guthrie to Robert Peel in 1829. 'Now the desire for knowledge induces many to overcome their prejudices and not only look at the dead body, but to hunt it out in the dissection room, and to examine all the bumps on the head, and compare the resemblance to the penny woodcuts.'[31] It is difficult to say how often this 'inconvenience' occurred, but at least surgeons might be thankful that phrenology had helped to change public feeling on the still delicate subject of human dissection.

Cerebral research promised a goal worthy of the medical profession. No less than anyone else, physicians desired a systematic mental science. The attempt by Dr John Kidd, regius professor of medicine at Oxford in the 1820s, to connect comparative anatomy to natural theology and other philosophic studies was not unique. Basically the profession welcomed the support which phrenology gave to the theory that the brain was the seat of the mind, for as late as 1820 this notion was far from general acceptance. The fact that it was so central to phrenology sometimes caused surgeons to play the devil's advocate and defend the new science. As for the localised mental functions, it is true that few in the profession thought there was adequate proof for the correlation between the skull and the organs, but that did not rule out the possibility that the brain had different physiological functions. And who, particularly among the surgeons, was to say that cerebral dissection did not contain an eventual clue? At a time when everyone else (including clergymen) was exploring the extent of the 'mental powers' the medical profession could hardly afford to leave any stone unturned if the mysteries of the cerebrum were to be conquered.

While surgeons differed among themselves about the meaning of the bumps, it was virtually impossible for them to disprove many of phrenology's answers. One such answer, popular to early psychology, was the explanation of the mind at rest. In 1830 Robert Macnish published a book which proved immediately useful to his fellow phrenologists and to his fellow physicians. There was in the *Philosophy of Sleep* a great deal which seemed reasonable to all: Macnish dealt with the effects of different temperatures and of oxygen, bedding, food and anxiety upon the sleep of men and animals; it was a subject on which phrenologists and physicians were able to exchange ideas profitably and openly.[32] Medical men found no real problem in accepting Macnish's theory of 'incomplete sleep', or the supposed activity of one or more mental organs while the others were at rest, as long as they were permitted to speak of functions rather than of precisely determined faculties. To some, the phrenological analysis of dreams and nightmares was similarly useful: phrenology made a connection between those who displayed an 'active *combativeness*' or high temper during the day and those who 'dreamed of broils and battles' at night.[33]

The *Philosophy of Sleep* suggested another concern fundamental both to phrenology and the practice of medicine. Many able physiologists were early attracted to phrenology not only for its emphasis on anatomy and dissection but also for its insistence on the conditions of good health and regular exercise. Apart from Robert Macnish, there was Dr John Elliotson of London, who translated Blumbach's *Physiology* and was a specialist in heart diseases. Elliotson (1791–1868), who delighted phrenologists in the 1820s and embarrassed them thereafter, was a fellow of the Royal Society and President of the Royal Medical and Chirurgical Society; as one of the young men who admired Spurzheim's surgical methods, he turned to phrenology about 1820. His fame as a phrenologist was, in contrast to George Combe, based on his career as a popular lecturer rather than a writer. Never was he more content than before a large audience, speaking of the merits of phrenology or demonstrating the techniques of mesmerism. Elliotson strongly believed in medicine and good health as a rational, commonsense business; for him phrenology removed much of the mystery attached to the practice of medicine. He admired Edward T. Craig, the Manchester phrenologist who preached the virtues of fresh air and open windows (a social cause soon to be known as 'Ventilation') and he respected Robert Macnish, who explained health in terms of proper rest and personal hygiene. These and other phrenologists were instrumental in making popular topics out of physiology and the laws of health, and Elliotson came to share their assumption that mental health, too, benefitted by exercise, clean living and a prudent diet.

No one made the connection between phrenology and physiology more effectively than did Andrew Combe. He never doubted that a philosophy of mind held the key to a total physiology of man; nor did he doubt that there was any variety of disease which was at least partly explained by phrenology.[34] Andrew Combe was known to talk about phrenology at the bedside of his patients. It was not phrenology in any great detail, with descriptions of faculties and nerves; instead it was the lesson of mind over matter, of understanding one's physical constitution and disabilities as the first step to better health. Dr Combe's counsel derived from personal recollections of poor food and clothing and of exposure to the cold and damp, as well as what he understood of the shape of the head. The phrenology of his *Physiology applied to Health and Education* (1834) was relatively unassuming; it certainly contrasted with the vigourous and didactic *credo* of his brother's *Elements*. Although Andrew believed as strongly as George in the reality of measurable mental powers, he tended to focus instead on the general role of the mind as an agent of activity. In Andrew's hands phrenology more closely resembled late nineteenth-century psychology; his *Physiology* was a similar synthesis of theory and observation, avoiding the dogmatism of hard and fast rules but still serving as a useful review for the physician and engaging study for the layman.

Despite his reputed shyness and modesty, Dr Combe was in a better position to advance phrenology among laymen than was Dr Epps (who spent his first year out of medical school lecturing on phrenology) or Dr Elliotson (who founded the London Phrenological Society). Dr Combe did not practice very long — he received his diploma in 1817 and retired in 1834 due to poor health — but his clients were remarkably well-placed. In Scotland they included newspaper editors and noble lords. In southern France and Italy, where he went to escape the rigours of Scottish winters, he also enjoyed many contacts. He early became acquainted with Sir James Clark, who introduced him to German princes. Eventually Combe was appointed physician to the King of Belgium — a lucrative and pleasant position which he relinquished only because of his own recurring illness. Returning to Edinburgh where he was a fellow of the Royal College of Physicians, Combe was made in 1838 one of the 'physicians extraordinary' to the Queen in Scotland, an office which he and his brother understood to denote 'professional respectability'.[35] While Combe owed none of these honours and distinctions to phrenology, it is true that phrenology benefitted from Combe's being so well placed.

The primitive state of medicine early in the nineteenth century helps us to understand why so many surgeons, anatomists and physicians thought phrenology true or partly true. The theory of the Four Humours was still widely believed and the cure of blood-letting was still widely practiced. Doctors were able to diagnose fairly well, but resorted to bizarre recipes to provide the cure. Dissection was now fully equated with anatomical research, but as the rise of phrenology indicates, it still allowed for a considerable amount of private judgment. There was little uniformity in medical training except its brevity: the most trusted physicians were very young when they received their diplomas — Sir

James Clark and Andrew Combe at the age of twenty-one, and John Epps at the age of twenty. The teaching of medicine was organised 'through private enterprize'; individual physicians established 'sketchy curricula' at the medical schools and often made fortunes in training new recruits.[36] Moreover there was no real pre-diploma practice or supervised internship. At the University College John Elliotson was supposed to have given patients large doses of drugs which his senior colleagues regarded as poisonous. There was, in the practice of medicine, an alarming similarity with school teachers of the day, in that the bulk of both professions often knew little more than those whom they were supposed to help.

It is not surprising, then, that phrenology made sense to many physicians, while others rejected its arguments not because they were able to disprove them but because of insufficient evidence. Even so, ignorance and inexperience were not the only reasons for their adoption of phrenology. The profession recognised the need to reform and modernise. The 'great object' which concerned Sir James Clark, Dr Richard Carmichael and others was 'to make medical men thinking animals' through better education, thereby raising the standards of practice.[37] It was in this quest that the profession turned to phrenology for ideas. Something had to be found to change the nature of medical thinking and Sir James Clark, for one, was always prepared to believe that a single revolutionary discovery would transform the whole profession. Like many doctors of the day he developed his own panacea. In a variety of tonic pills which he produced, Clark claimed to have come upon an infallible cure for indigestion, heartburn, nausea and 'general debility'; the medicine was guaranteed and even Her Majesty is said to have found it useful.[38] But Clark realised the shortcomings of tonics and acknowledged the 'unquestionable truth' that mental disorders called for a different sort of treatment — assuming all the while that the mind was as treatable as the body. Like his friend John Conolly, Clark thought it impossible to overlook the dimensions of a patient's head whenever there was a question of recovery from illness; the faculty of *conscientiousness* revealed whether the patient was likely to follow the doctor's orders.

John Conolly (1794–1866) was particularly intrested in the role of the mind in dealing with a variety of physical ailments. In effect, he expected that all future doctors would have to be excellent psychologists. The very word psychology, only now coming into the language, was by the 1850s commonly associated with the work of phrenologists; for Clark and Conolly it replaced older expressions like 'mental science'. Conolly was afraid (as, to a lesser extent, was Clark) that medicine was becoming relatively backward compared to other professions; how was it to be brought up to date? For some of his colleagues the answer lay in mesmerism or homoeopathy, but Conolly was satisfied with the phrenology of Spurzheim. As a young medical student Conolly heard some of Spurzheim's early lectures and he was impressed by the doctor's knowledge of anatomy. He was also interested that Spurzheim should confront the problems of insanity, of the 'imbalanced mind' and of states of 'melancholia' in otherwise healthy persons: it was the very field to which Conolly devoted his career. He saw phrenology as a dynamic study of human powers and as the only convincing explanation of the disordered mind. As a tribute to phrenology Conolly founded the local societies in Warwick and Leamington, but it was not remembered that phrenology was the inspiration behind his work in the asylums and hospitals of the Midlands.[39]

Others of his profession reinforced phrenology's wisdom with that of another fringe science. The benefits of 'phreno-mesmerism' (as Dr Caldwell first called it) seemed more immediate, especially to the practice of medicine. In England Harriet Martineau, only recently converted to phrenology after a long flirtation with the science, was converted in turn to the new amalgam of phrenology, mesmerism and clairvoyancy. She insisted that it cured her of illness and fatigue and — lest George Combe have any doubt about it — she affirmed that she had been mesmerised by 'a very good phrenologist'.[40] It was inevitable that phreno-mesmerism should have arisen. Phrenology paved the way by proving the existence of mental faculties, and all that remained was to show that an independent force might activate any of the faculties. Those which received sense perceptions — *colour, form, tune* and the like — were particularly susceptible to mesmerism. The dynamics of the union were explained by M. B. Sampson in a letter to Combe. A phrenologist of long standing, Sampson was invited to a mesmerising experiment and was asked to touch the head of a person who had apparently been induced into a deep trance. Resting his finger on the faculty of *tune*, Sampson was surprised to hear the person ask questions about musical instruments and then begin singing.[41] Combe was naturally reluctant to believe the story and, when he had the opportunity two months later, he attended a similar meeting in Edinburgh. Writing about the event several days afterward, Combe was still astonished, for the demonstration produced the same results as Sampson had described from London. The subject made 'very appropriate remarks', though not in full sentences, whenever various faculties were touched. Combe insisted on touching the head himself to be certain there was no trickery; he also ascertained that the person knew nothing about phrenology. Even so, each touch brought a suitable comment.[42]

It is not difficult to understand why the mesmeric reformation of phrenology gained favour among educated people. Both sciences claimed a foundation of observable phenomena, while at the same time both probed the unknown. To treat an audience to a demonstration of thought transference or clairvoyancy (both associated with phreno-mesmerism) was like conducting them on a tour of an unknown land. Mesmerism was the forcible opening of undiscovered territory, the inscrutable frontier. Dr Epps and Dr Elliotson thought phreno-mesmerism the only way of eliciting information from difficult patients or of determining the cause of mental depression; it was also useful in detecting signs of mental disease which the conscious patient might well disguise. Phreno-mesmerism was an ambitious business, too, as it sought to explain (and promote) thought-transference. Sampson thought the process worked like the new electric telegraph which transmitted up to ninety letters per minute. The assumption was that Nature must have some way of showing what was in the air but could not be seen. The electric telegraph held the promise of swifter communication in the future; and were the powers of the mind to play their role at last? Now there was the use of ether to produce the proper state of mind, and phreno-mesmerists used the liquid in order to eliminate the charges (so often made in the past) of collusion between operator and patient. Some physicians adopted phreno-mesmerism because they were able to defend it as the forerunner of modern anaesthesia, and it was not long before the public associated phreno-mesmerism with such innovations as ether and the stethoscope.[43] In fact one could hardly avoid making the connection when, in 1849, Dr Elliotson opened his 'mesmeric hospital'.

And so 'craniology' became, as its opponents feared, a professional issue and, very largely, a professional affair. Given the general desire among medical men for a breakthrough in knowledge of the human condition, it is not surprising that many of them should have embraced phrenology. *The Times* and the Edinburgh *Review* could rant all they wished, but Spurzheim made his impression and won disciples among the doctors of Britain. The *Gentleman's Magazine* was probably closer to the truth when it reported Spurzheim's success at a London lecture: 'The professional gentlemen present (being all the best anatomists and most distinguished physicians in the metropolis) admitted the justness and originality of the professor's observations.'[44] Of course the affiliation was never exclusive. Professional scientists as well as physicians and surgeons were attracted to Spurzheim's gospel. Phrenology was, after all, definable as both a philosophy and a science, and as a group phrenologists were very enthusiastic about all forms of scientific inquiry. In fact, they never ceased singing the praise of science. In their correspondence phrenologists were always urging one another to study botany, chemistry or astronomy; later editions of the *Constitution of Man* provided a definition of geology (which Combe spelled with a capital 'g') and quotations from contemporary geologists. Everyone was fascinated by the study of natural history in the 1830s and forties, and the phrenologists were no exception. They were to be found among the gentlemen geologists who scoured the countryside on weekend expeditions, searching for curious shells and strata. Phrenologists made the lessons of geology, chemistry and natural history their own, and certainly one reason for the popularity of the *Constitution of Man* was the author's talent for writing about the 'balance of creation'.[45] His was not a shrewd design to exploit the current fashion or to make the most of a middle-class enthusiasm, for phrenologists were quite genuine in their tributes to science. And while scientists did not return the compliments to the same extent as did the medical profession, they were by no means all opposed to the doctrines of Gall and Spurzheim.

In Scotland one of phrenology's most promising and obliging defenders was the geologist and mineralogist Sir George Stewart Mackenzie. A member of the Royal Societies of Edinburgh and London, Mackenzie travelled widely and in 1811 published a report showing the geological similarities of Scotland, Iceland and the Faroe Islands. Turning to phrenology about 1818, Mackenzie felt that only a bold and continuous barrage of presentations and propaganda would ever convince his colleagues of the truth of phrenology. He therefore decided to stir them up himself. Early in 1830 he raised the subject of phrenology before the Royal Society of Edinburgh where undoubtedly many of the members, still upset by the Spurzheim-Hamilton debate of two years earlier, regarded phrenology as a persistant nuisance. In his address Mackenzie said that he could not believe that the Royal Society of Edinburgh would remain hostile to the new science. True, he himself used to condemn it (in his candour Mackenzie often spoke like Paul confessing to a role in persecutions), but was it too much to hope that the members would take the time to investigate phrenology, as he had done? Mackenzie then summarised the merits of his mental science and proclaimed Gall as possibly the greatest modern philosopher and benefactor of mankind. The Society (whose meeting was chaired by the unbelieving Sir Walter Scott) was not agreeable to the idea of investigation, but Mackenzie was not discouraged. On an individual and personal basis he continued to suggest that scientists and writers, who prided themselves as impartial investigators, give phrenology a fair hearing.

Mackenzie's task was just as hard, if not harder, than trying to make converts of the medical profession. The members of the Royal Society were reluctant to entertain discussions of phrenology because of its implications and applications. In 1834, for instance, the 'statistical section' of the British Association for the Advancement of Science sent a reminder to the Association as a whole that it was unwise to introduce 'moral and political questions' or to 'overstep the prescribed boundaries of the Institution'. The warning did not specify the persons or ideas which might lead the Association astray, but the *Phrenological Journal* had no doubt that phrenologists were the object of attack.[46] Phrenology continued nevertheless to find its way into such learned councils and in 1834 (probably much to the annoyance of the statistical section) membership in the British Association was extended to George Combe.

For the natural scientist, the most interesting aspect of phrenology was its trust in the continuing development of man and creation. Phrenologists suspected that no part of the physical universe was changeless — least of all the human brain — and in the *Constitution of Man* Combe spoke of a world arranged according to a principle of gradual and progressive improvement — a principle so wide in its application as to suggest the theory of evolution. Combe noted that creation had endured many revolutions in the 'preparation of something better' and that there had been 'successive orders of living beings, rising higher and higher in the scale of intelligence until man appeared'.[47] All phrenologists were intrigued by evolutionary theory and they were fascinated by the cerebral consequences of 'generations of warfare by successive inhabitants of our planet'. The debt to evolutionists was a happy one in that phrenologists always acknowledged it honestly. At the same time their role was not altogether unoriginal. They contributed to the theory by insisting on the likelihood of mental evolution, and they thought the Mosaic estimate of creation unacceptable for the length of time really needed to produce man's superior faculties. How then did man fit into creation?

This is where the phrenologists, and later the Darwinists, ran into trouble. They claimed to be content with the meaning which enlightened theologians attached to the phrase 'In the beginning', but many phrenologists suspected that man really owed his origin to an unknown amphibian. It was not an idea they cared to publish; it was instead confined to their correspondence. Even after the appearance of Darwin's *Origin* in 1859, British phrenologists were very reluctant to circulate the sort of evolutionary charts which phrenologists on the continent were publishing. Typical of these bold expressions of evolution was the *Eléments de Philosophie phrénologique* (1861) by Henri Scoutetten. His booklet included a fold-out page which showed the progressive development of man in twentyfour heads – beginning with that of a tadpole and culminating in the noble visage of Apollo Belvedere. British phrenologists did not want anything quite as assertive, and they cautiously expressed in words what they were afraid to illustrate with engravings. In any event, it was not a very great step from the phrenology of the 1820s to the speculative evolution of the fifties and sixties; the two main phrenological journals (one in Edinburgh and the other in London) frequently described the various species of animals, calling attention, as did the *Constitution of Man*, to the similar faculties of human and animal brains and hinting at the possibility of historic progression. In this fashion British phrenologists were not merely advertisers of evolutionary theory; in the persons of Sir George Mackenzie, Hewett Watson and Robert Chambers, they actually helped to produce it.

It was not absolutely necessary to trace the development of man or the transformation of Nature by reference to skulls. Geological studies such as Mackenzie conducted added to the argument that Genesis could not be taken literally, and Hewett Watson supported the same conclusion through botanical research. Perhaps Watson's contribution was even more important. Having studied law and medicine at Edinburgh before turning to botany, Hewett Cottrell Watson (1804–81) took up phrenology as a philosophical adjunct to his scientific investigations. The *Statistics of Phrenology* and the *London Catalogue of British Plants* were the products of Watson's plodding scholarship but as a thinker he was not entirely unspeculative. To Watson it appeared that Nature's 'earlier species' were of relatively simple organisation and were gradually replaced by what he called 'fresh species'. From botanical evolution it was but a short distance to human evolution and, as Watson wrote to Combe in 1847, the race of man was only 'the latest created species . . . immeasurably beyond any other in mental endowments'.[48] Generally the sacrifice of the earlier species was the sacrifice of the inferior for the sake of superior creatures, and it was not inconceivable that man himself might someday be surpassed or replaced. More than a decade later, Darwin was to enunciate what Watson and other naturalists had been mumbling for some time, and after reading the *Origin of the Species* Watson wrote a long and interesting letter to the author. With it he enclosed a copy of his defence of the *Constitution of Man* and asked Darwin to notice the markings in the margins. 'You will discover', Watson wrote rather proudly, 'that a quarter of a century ago, I was one of the few who then doubted the absolute distinctness of species, and the several creations of them.'[49]

Darwin had no reason to doubt Watson's role in the controversy. Watson, after all, deserved every respect for his training as a professional scientist. What Darwin and others of his profession objected to was the attention given to evolution by non-scientists. And yet this latter class of writers (surprisingly numerous by the 1840s) performed an almost thankless job in raising the issue as the advance guard against religious opposition. The most notable (and successful) of this group, a phrenologist who was 'animated by a scientific purpose', was Robert Chambers. As a journalist and writer, Chambers (1802–71) was a director of the famous *Journal* which, with his brother, he established in 1832. He derived his phrenology from a series of lectures given by George Combe in Edinburgh, and his conversion to the science in 1834 was, by his own admission, very timely. As he later confided, the *Chambers Journal* (a weekly devoted to 'Literary and Scientific Subjects') had suffered from declining sales for almost a year. At this point Chambers became a phrenologist and decided to 'introduce covertly' the doctrines of the *Constitution of Man* into his *Journal*. 'From that moment', as Combe relayed the explanation to a mutual friend, 'the circulation began to rise, and it has risen 10,000 on each impression since that time.'[50] The explanation is hardly credible, although Chambers could not imagine any cause for this achievement other than the 'superior sense and utility' of phrenology. There was actually no great change in the format of the *Journal*; the selection of biographies, travelogues, practical science and items of natural history all remained. If the *Constitution of Man* had inspired Chambers in any way, it was as an example of how various sciences might be amalgamated and dressed up for popular consumption. The *Journal* worked as a synthesis of many things, and his accomplishment encouraged Chambers (amateur as he confessed himself to be) to enter the field of scientific controversy.

In 1844, after two quiet years of work at his St Andrews retreat, Robert Chambers published the *Vestiges of the Natural History of Creation*. This work, which Darwin thought 'strange, unphilosophical but capitally written',[51] provoked an immediate uproar because of its unorthodox treatment of the Creation. Chambers anticipated this reaction and therefore arranged for the book's publication in Manchester under the direction of his friend (and fellow phrenologist) Alexander Ireland. The *Vestiges* appeared anonymously, and it was not until 1884 that Ireland officially revealed that Chambers had written it. For years the authorship was a subject of popular debate, although George and Andrew Combe knew the secret almost from the beginning. It is interesting that while Darwin and other professional scientists felt that the *Vestiges* was a crude and illogical approach to evolution, others thought it sufficiently respectable to have been written by a member of the intellectual community; the guesses included William Thackeray, Sir Charles Lyell, Richard Vyvyan, George Combe and even Prince Albert.

The one certainty about the author was his commitment to phrenology. He spoke of the mental constitution of men and animals and reverently quoted Dr Gall.[52] He properly defined *cautiousness* and *secretiveness* and he declared that every individual was merely part of 'an intensive social mechanism'.[53] Admittedly Chambers' use of 'comparative phrenology' was not the best way to account for the intellectual and moral qualities of different people;[54] but as an answer it satisfied Chambers, and the immense popularity of the *Vestiges* (the first ten editions selling close to a quarter million copies) indicated that the public was satisfied too. Chambers and Combe thought that the reception of the *Vestiges* meant that the whole issue of man's origin was an open question for almost everyone, and they were delighted that the 'evangelical hostility' was not as ferocious as it might have been.[55] Pleased by the phrenological approach of the *Vestiges* Combe urged the author to publish a cheaper edition for libraries. It was good advice and Chambers took it. Under the circumstances he almost felt compelled to do so. For just as he had unknowingly helped to cushion the shock which accompanied Darwin's *Origin*, so too the *Constitution of Man* paved the way for the *Vestiges of Creation*. There were many, in fact, who felt justified in seeing an anti-religious conspiracy; they were now quite certain that Combe's great essay of 1828 was consciously designed as a prelude to the evolutionary ideas of the *Vestiges*.

George Combe never actually pledged himself to the notion of human ascendency from a primeval or amphibian animal. Yet in spite of his caution, he was charged with harbouring the most extreme views on the subject. In the 1830s Watson came to his defence by pointing out that evolution in the realms of geology and botany did not necessarily mean evolution in the human species — although privately many phrenologists thought it did. In Watson's words, Combe was 'only inclined' to evolution.[56] But careful words were not enough to close the gap which Combe had already acknowledged in the *Constitution of Man* between orthodoxy and evolution. On one side of the intellectual chasm, man took his place at the head of all other creatures, his brain 'unquestionably the workmanship of God',[57] But the rest of Combe's analysis went in the opposite direction. He considered man a 'progressive being' left to his own devices and subject to natural laws which influenced everything else. And what exactly did he mean by progression? Was man in time to become a different creature? To such questions the only reply Combe gave was that of the rationalist: he insisted that 'History exhibits the human race only in a state of progress towards the full development of its powers, and the attainment of rational enjoyment'.[58] Taken separately the words were not terribly offensive, but there was nothing tentative about the whole answer. It appeared time after time, in every edition of the *Constitution of Man*, and doubtless it was still fresh in the minds of many anti-evolutionists who, in 1844, fell upon *Vestiges of Creation*.

In contributing to the storm over evolution, phrenologists were in a strong position. Indeed no single issue (save for the narrower one of racialism) was to benefit them more. The popularity of the *Vestiges* certainly indicates that the public was interested in the problem and decidedly receptive to a phrenological explanation of it. The fashionable deism of phrenologists left just enough room for the Almighty to manoeuvre and, this being the case, respectable opinion was not scared away. Nor could any intelligent gentleman easily ignore the connection which phrenologists made between the evolution (or development) of Creation and the whole body of natural law. Those who neglected the natural laws, or attempted to circumvent them, did not often survive; and if they did remain, they rarely achieved pre-eminence among members of their species. In the *Constitution of Man*, Combe announced that his object was to prove that 'rewards and punishments of human actions are infinitely more complete, more certain, and more efficacious in this life, than is generally believed'.[59] Cast in the blinding light of the natural laws, the concept of evolution was not so implausible. James Simpson, one of the leading Scottish phrenologists, once reported that among his fellow lawyers phrenology was attractive for this reason: it emphasised human conduct, human intelligence, and human responsibility; it rendered the most heinous crimes explicable while it upheld all those laws and relationships which man ignored at his peril. For men of Simpson's professional calibre, phrenology was thus comforting and reasonable. As respected men in their society, they must have been gratified to learn that the causes and motives of human activity were as knowable as the human laws which governed society. To gentlemen whose duty it was to protect the righteous and punish the wrong, the *Constitution of Man* sounded the right note. The doctrines of phrenology (which the *Gentleman's Magazine* at first glance condemned) proved upon investigation to be 'neither fanciful nor dangerous to the intellectual part of mankind' but worthy of all men 'who desired to be acquainted with the Philosophy of Mind'.[60]

Notes

1. *Discussion on Phrenology; between Charles Donovan and the Reverend Brewin Grant* (London, 1850), pp.6—10.
2. *Philosophical Transactions of the Royal Society of London* (London, 1823), CVIII, p.306; also in *Nervous System of the Human Body* (London, 1830), p.222.
3. Macnish, *An Introduction to Phrenology, in the form of Question and Answer* (Glasgow, 1836), pp.9 and 163f.
4. Martineau, *Biographical Sketches*, p.273.
5. Holyoake, *Sixty Years of an Agitator's Life*, I, p.60.
6. Combe, *Essays on Phrenology*, p.5.
7. *Encyclopedia Britannica* (7th ed.: Edinburgh, 1842), XVII, pp.454—73.
8. U. C. Knoepflmacher, 'The Use of Classification: "The Wellesley Index" ', *Victorian Studies*, X, no.3 (1967), p.263.
9. Combe, *Elements*, p.20.
10. Reid, *Essays on the Intellectual Powers of Man* (Edinburgh, 1785), Essay II, p.96.
11. George Combe, *Lectures on Popular Education; delivered to the Edinburgh Philosophical Association* . . . (2nd ed.: Edinburgh, 1837), p.103.
12. Combe, *Constitution of Man* (5th ed.), p.11.
13. George Combe, *Notes on the new Reformation in Germany, and on National Education, and on the Common Schools of Massachusetts* (Edinburgh, 1845), p.3f.
14. John Epps, *Horae Phrenologicae; being three phrenological Essays* (2nd ed.: London, 1834), pp.20—3.
15. Combe, *Constitution of Man* (1st ed.), p.53.
16. *Ibid.*, pp.28—9.

17. Macnab, *Analysis and Analogy recommended as the Means of rendering Experience and Observation useful in Education* (Paris, 1818), p.25.
18. *Ibid.*, pp.120–1.
19. *Journal*, I, no.4 (1824), p.622.
20. Epps, *Horae Phrenologicae*, pp.52, 68–70.
21. *Ibid.*, p.35.
22. Edinburgh *Review*, XXV, no.49 (June 1815), p.254.
23. H. M. Knox, *Two Hundred and Fifty Years of Scottish Education, 1696–1946* (Edinburgh, 1953), p.20.
24. Handbill in Combe Papers, NLS 7292/24.
25. *Britannica, op.cit.*
26. Watson, *Statistics*, p.129.
27. For the phrenologists' view of divisions within the medical profession, see *Journal*, XIII, no.53 (1840), pp.128–41.
28. *Medico-Chirurgical Review*, 28 March 1825.
29. The *Scotsman*, 26 January 1828.
30. *Caledonian Mercury*, 28 January 1828; T. S. Baynes, 'Sir William Hamilton', in *Edinburgh Essays, by Members of the University* (Edinburgh, 1856), p.261.
31. Dr G. J. Guthrie, *Letter to the Rt. Hon. the Secretary of State for the Home Department, containing Remarks on the Report of the Select Committee...* (London, 1829), p.16.
32. One instance of cooperation was the publication of Dr Lyon Playfair's article on 'Sleep and some of its concomitant Phenomena' in the *Journal*, XVII, no.81 (1844), pp.327–42.
33. *Journal*, I, no.3 (1824), p.354; also Henry Turner, *Phrenology: its Evidences and Inferences, with Criticisms upon Mr Grant's recent Lectures* (London, 1858), p.4f.
34. Andrew Combe, *Phrenology – its Nature and Uses: an Address to the Students of Anderson's University* (Edinburgh, 1846), p.12.
35. George Combe to Caldwell, 29 August 1838. NLS 7388/84.
36. Harley Williams, in his account of the life of John Elliotson, in *Doctors Differ: Five Studies in Contrast* (London, 1947), p.27f. The contemporary approach to anatomy is illustrated in various publications, including Frederick Tyrrel, *Introductory Lecture on Anatomy; delivered at the New Medical School, Aldersgate Street...* (London, 1826), p.26 *et passim*.
37. Clark to Combe, 7 April 1843. NLS 7267/66.
38. A clipping from the *Scotsman*, advertising Clark's 'Tonic Pills', in the Combe Papers, NLS 7313/130.
39. The *Dictionary of National Biography* never mentions phrenology in its article on Conolly. The omission is typical of many late nineteenth century biographies.
40. Martineau to Combe, 27 September 1844. NLS 7273/64.
41. Sampson to Combe, 25 March 1843. NLS 7270/18–19.
42. George to Andrew Combe, 23 May 1843. NLS 7380/25.
43. In his practice Elliotson was reputed to be among the very first to use these innovations. Another name for Caldwell's phreno-

mesmerism in America was 'etherology'.
44. *Gentleman's Magazine*, LXXXIV (1814), p.525.
45 *Constitution of Man* (5th ed.), pp.3—6; also *Lectures on Popular Education*, p.7.
46. *Journal*, IX, no.42 (1834), p.121.
47. *Constitution of Man* (5th ed.), p.3.
48. Watson to Combe, 14 May 1847. NLS 7288/126.
49. *The Life and Letters of Charles Darwin, including an Autobiographical Chapter*, ed. Francis Darwin, (3 vols., London, 1887), II, p.227.
50. Combe to Alexander Maconochie, 6 July 1835. NLS 7386/347.
51. *Life and Letters of Darwin*, I, p.333n.
52. [Robert Chambers] *Vestiges of the Natural History of Creation* (2nd ed.: London, 1844), p.324 *et passim*.
53. *Ibid.*, p.355.
54. Charles Gillispie, *Genesis and Geology: a Study in the Relations of Scientific Thought, Natural Theology and Social Opinion in Great Britain, 1790—1850* (New York, 1959), p.156f.
55. Chambers to Combe, 14 March 1847. NLS 7283/127.
56. H. C. Watson, *Examination of Mr Scott's Attack upon Mr Combe's 'Constitution of Man'* (London, 1836), p.20f.
57. *Constitution of Man* (5th ed.), p.5.
58. *Ibid.*, p.10.
59. *Ibid.*, p.22.
60. *Gentleman's Magazine*, LXXXV (1815), p.441.

Chapter IV A Shortcut to Knowledge

There was a more mundane side to the spread of phrenology. Not everyone who read the *Constitution of Man* was a rationalist or a deist or as spellbound as James Simpson was by the beauty of the laws of Nature. Nor did they all hope to apply phrenology to one or another of the learned professions. On the contrary, it was Hewett Watson's opinion in 1839 that phrenology was more appreciated by common people than by the 'superiorly trained intellects'.[1] Why was this? Partly it was the work of the money-minded amateurs: those who 'read heads for hay'. By 1840 they were to be found everywhere in Britain and America and their work, far from gaining Combe's *imprimatur*, was unlicensed and resented. Moreover they were thought to undermine phrenology as a philosophy and as a proper science by their dogmatic judgments in such fields as marriage counselling, portrait analysis and vocational guidance.

Not that the followers of Dr Spurzheim were disinterested in such fields. They became increasingly conscious of their role as popular philosophers and it was very difficult, even for the most orthodox among them, to neglect the market of popular head-reading. In its later days the *Phrenological Journal* devoted more space to the interpretation of mental powers when only pictures or statues were available, and in 1855 George Combe himself published *Phrenology applied to Painting and Sculpture*, which involved the sort of analysis which earlier phrenologists (including Combe) thought secondary to their calling. But to the general public these judgments were plausible because the phrenologists, often bearing credentials of professional men, were able to weave together what was scientifically known with what was quite unprovable. In Edinburgh this merger occurred in the lectures given under the auspices of the Association for Providing Instruction in Useful and Entertaining Sciences. Open to the public, the lectures were arranged under the broad headings of Chemistry, Geology, Natural Philosophy and Physiology; the science of Gall and Spurzheim made the list in its own right. The divisions, however, were not very precise. The phrenologists, who were able to speak about the more traditional sciences as well as their own, were often invited to do so; the result was that, in the course of one evening, their audience imbibed the wonders

A Shortcut to Knowledge 59

of biology, the principles of physiology, and the mysteries of the brain.

No-one thought it necessary or advisable to apologise for this blending of knowledge. And countless audiences, composed of people of every trade and born of a generation which believed in the power of knowledge, never objected. Even those sceptical of phrenology itself thought the lecture series was (in the words of the advertisement) both useful and entertaining. One anonymous ticket-holder, signing his name 'Candidus', derived so much satisfaction that his basic doubts about phrenology did not matter. The important thing, he told Combe, was to 'dispel the profound ignorance among all classes'.[2]

There was Hewett Watson's answer. As a professional scientist (and as a pioneer of the public opinion poll) he ought to have realised that phrenology offered greater advantage to the beginner rather than to the master of studies. He should have realised too that among ordinary people the most alluring feature of phrenology was its apparent synthesis of scientific information and general wisdom. It was this fascinating *résumé*, accomplished by a plain description of the human condition, which delighted the youngest and the least tutored intellect. What other philosophy, traditional or scientific, bothered to talk about planets and fishes, or visited primitive peoples, or referred to the great beasts which preceded the human species? The *Constitution of Man* embraced such topics as part of a popular history of mankind. It was easy information and surprisingly pertinent: in discussing the merits of phrenology, the last chapter of the *Constitution of Man* treated the problems of crime, education, religious observance and good government, and while the remarks were obviously fleeting in nature they were nonetheless right up to date. At a time when knowledge was both power and powerful entertainment, no-one could expect more than this.

There is little wonder, then, that phrenologists were looked upon as trustworthy purveyors of knowledge. Some of the knowledge was patently scientific, some of it philosophical, and some psychological. The bulk of it seemed practical, inter-locking and easily digested by the common folk who came in great numbers to hear George Combe, James Simpson or John Epps. In one sense, the composition of these audiences suggests what they were hoping to find. Most of the professional people were usually medical men of some description (whose interest in phrenology we have already encountered) or teachers (who will, as a group, concern us elsewhere). But shopkeepers came too, and so did artisans and housewives. Apart from the phrenology, which they may or may not have accepted, they were looking for a brief and entertaining introduction to the world of knowledge, a glimpse at the information which their rank or their lack of leisure time may have denied them. It was for this vast portion of society that Combe prepared in 1836 a 'people's edition' of the *Constitution of Man* which found its way into so many of the mechanics' libraries and working-class homes of Britain.

The response was predictable. The people's edition sold 17,000 copies in little more than a year partly because it was a short cut to useful information, partly because the price (11s. 6d.) was right, but primarily because it outlined a whole philosophy 'for ordinary people [who] for the first time have ground whereon to rest the soles of their feet'.[3] The ground was solid indeed because, as we have seen, phrenology abstained from metaphysics and dealt with observable phenomena. Yet it was not merely a variety of positivism. Phrenology became almost a way of thinking because it took the old virtues of life and demonstrated in a novel way how each was necessary, natural and attainable. Self-knowledge, for example, was necessary if man as a rational creature was to serve his community and lead a useful life. Temperance was a natural course of action if he wished to preserve the freedom of his own mental faculties and the mental health of his children. Honest work satisfied the dictates of *conscientiousness*. Keeping the Sabbath fulfilled the impulse of *veneration* and gave the mind time for rest. Such were the virtues of a respectable life which Combe incorporated into the *Constitution of Man*. They were borrowed virtues, to be sure, but they were defined and exhalted, and each was given a practical significance.

Of these virtues, self-knowledge was crucial. And it was more easily attainable than the philosophers of old ever imagined. Using phrenology, one was now able to appreciate his own distinctive personality and to find out which of the mental faculties most influenced his conduct. Assuming that man was a rational creature and 'not merely a tame beast', he should be able to know his own talents and shortcomings. Self-discovery was also essential if man was to fulfill his spiritual obligations. As the author of *Christian Phrenology* asserted, the virtuous life was a state of constant trial, and no man would triumph unless he were completely honest with himself.[4] To that extent, the process of self-knowledge was not in itself the goal: it was only the basis for self-improvement. Man could hardly expect to employ his mental resources properly if he did not know what they were.

That was the crux of the message which phrenologists brought to the mechanics' institutes. Of course they encouraged artisans to learn the principles of political economy and something of the natural sciences; such lessons were important in an expanding industrial society. But in their opinion it was far more important to learn about human physiology and the benefits of mental health. Joshua Toulmin Smith, a familiar figure at mechanics' institutes, argued the case for physical and mental exercise; both, he said, were necessary for the self-improvement of the individual and for a stable and happy society.[5] J. L. Levison presented a similar argument. He told mechanics what he had already proclaimed in his book on *Mental Culture*: the faculties of the mind, like the limbs of the body, were weakened by the lack of exercise; the 'natural means of improving the human race' was to stimulate the moral and reflective organs of the mind.[6] Still more explicit was Dr John Abercrombie, a patron of mechanics' libraries in Scotland. He spoke to his audiences about the 'regulated condition of the mind'; he believed that in most sane and sensible persons the organ of the mind was understandable and manageable.[7] In other words, phrenology's advice was not unlike the gospel story of the three men with different talents. Not everyone was born (or entrusted) with the same abilities, but man owed it to himself to make the most of the talents he had. According to Henry Turner (another lecturer popular among artisans), only phrenology, a 'systematic arrangement of the facts', helped man to obey the injunction 'know thyself'.[8] Independently of social and political change, working-class audiences were told they might secure a better life if they were always directed by self-knowledge and inspired by mental discipline.

Mental discipline implied the 'positive exercise' (or exercise to the good) of each faculty of mind. While phrenologists were careful not to promise miracles by this process, there were nonetheless a few cases of real attainment. When speaking to working-class groups, Dr Epps liked to recall that when he was first received as a member of the Edinburgh Phrenological Society he was advised to cultivate his perceptive powers (already fairly strong) by the 'study of individual objects'. Epps dutifully followed this counsel by studying natural history and botany. He was delighted with the results. Not only did the self-discipline strengthen his mind: it also brought him a gold medal for his proficiency in botany at Edinburgh.[9] As in the case of Samuel Smiles's edifying biographies, the moral was clear: what Dr Epps had done, others could do.

To improve one's mental condition, whether marginally or as remarkably as Dr Epps claimed to do, a man had to be phrenologically examined and given a prescription to meet any deficiency. If, for example, a builder never did his share of work, the phrenologists might hope to stimulate his faculties of *wonder* and *conscientiousness* by showing him what fine houses other men had constructed and how their work had won security for themselves and their families. To this goal, the builder's reading material, as well as his recreations and pastimes, should be calculated to encourage the proper social attitudes and mental dispositions. This is why George Combe (not alone among phrenologists) thought so little of novels and other 'soothing, sentimental material' which exercised only the feelings. Persons of good sense and social conscience read books 'related to human nature, life and to one's duties'.[10] Phrenologists were inclined to be earnest people who cared little for literature that ignored the 'mental history' of its characters or the deep-seated reasons for their conduct.

There was one very obvious limit to mental self-improvement. While mental deficiencies could be identified and while some effort could be made to rise above them, they could never be eliminated completely. The best one might do was to restrain or redirect certain tendencies. Within these general bounds, however, men of every rank were able to 'shun intemperence and all pernicious habits'. A knowledge of personal weaknesses was the first step towards counteracting them and reducing the inconvenience which they caused to others. Knowing what they did of human nature, the phrenologists realised that the complete reversal of mental tendencies was an illusion, given the short space of a single lifetime. They also recognised that some 'pernicious habits' were very difficult to control — let alone irradicate. Of these, intemperance was the most formidable, and phrenologists regarded it as the greatest danger both to mental health and to social responsibility.

They did not look upon drink in quite the same way as most temperance crusaders did. Intemperance was an affront to Reason rather than a sin, and in advanced cases it was a form of mental derangement. No one ever discovered a faculty responsible for the love of intoxicating drink, but there was no doubt that chronic drunkenness was the result of innate mental conditions.[11] Phrenologists claimed (and many non-phrenologist surgeons believed) that the dissection of the brain revealed these abnormal conditions. Dr Andrew Combe and Dr Charles Caldwell frequently drew public attention to the diseased brains of criminals notorious for their fondness of alcohol. As educators too, phrenologists were sensitive to the social effects of alcoholism. In the *Philosophy of Education*, James Simpson expressed the opinion (held in many quarters) that intemperance was a hereditary quality in families, 'breaking out at the same age in several individuals of the same stock'.[12] J. L. Levison took the same view in a work which he dedicated to Dr Birkbeck, the founder of the London Mechanics' Institute. Parents must realise, Levison warned, that the physical and moral sins of fathers were visited in their children, and that all men become drunkards first by 'being moderate'.[13]

Levison was more vehement on the subject than most of his phrenologist friends. Without compromise he advocated a hard line on drunkenness. So stringent was his argument (much to the embarrassment of Combe and Simpson) that he antagonised members of the temperance movement who were otherwise sympathetic to phrenology. Levison mounted his attack not only against the excesses of drink but against drink itself, lashing out against the temperance societies which he called 'indirect seminaries of drunkenness' because they sometimes recommended wine and beer in place of ardent spirits.[14] The Combes tried to restrain Levison even though they admitted that there was just cause for alarm. The fact remained that there was a 'remarkable increase of insanity' among members of the lower orders, particularly in manufacturing districts, which the *Phrenological Journal* had already attributed to the practice of dram-drinking.[15] How was society to meet this evil? Obviously it was no good to persuade everyone to 'take the pledge', for the ignorant and depraved could hardly be expected to keep a vow. Simpson, while not agreeing with Levison, thought that many persons were indifferent to the cause of temperance simply because they were mentally incapable of understanding the mischief of drink. In those cases of an 'innate love of liquor' there was little hope of rescuing the drunkard, whose mental constitution consigned him to the lost. In those cases where there was yet hope, Dr Andrew Combe advised that every effort be made to treat the drunkard as a patient requiring moral as well as physical help.[16]

Phrenology's attitude to the social problem of drunkenness requires a note of caution. We must not assume from the rashness of Levison's career and from the dire warnings of other phrenologists that the fatalism of their science detracted greatly from its popularity. The fact is that phrenology inspired many of the rank and file of the temperance movement in Britain. They were impressed by its physical arguments and its concern for the health of the whole man; they were also comforted by its assurance that those lost to drink or hereditary insanity were few in number. Phrenology, while very gloomy on the prospects of saving the few, was absolutely cheerful on the potential progress of the vast bulk of the population. Fundamentally the British mind was sound, susceptible to progress and quick to improve. At home phrenology was an optimistic creed, and the uncounted millions of inferior minds elsewhere in the world were of small concern to British phrenologists. Self-knowledge and self-improvement had become the national goals of a people whose intellectual powers were second to none. Phrenologists were forever tracing the happy record of British achievements; they were, almost self-consciously, prophets of progress who understood their society in terms of mental prowess. 'As far as we can learn from historical records', wrote William Ellis at the time of the Great Exhibition, 'everywhere around us we see things better done than in former times [and] we learn nothing to shake this cheerful conviction or to mortify us that we are falling behind other nations.'[17] The products of British genius — the railroads, the efficient mills, the electric lights which M. B. Sampson saw in Trafalgar Square — were the work of quite ordinary people whose persistence and inventiveness had made them successful. The accomplishments of these men waited only upon the ingenuity of those to come. Such was the confidence in British self-improvement and in western civilisation generally that George Combe, in his book on the *Relation of Science and Religion*, asserted the idea of an advancing world, one instituted for a good end and not merely the wreck of a higher order of things.[18]

The optimism of phrenology, in other words, far outweighed its fatalism. Working-class audiences were told that they could learn, and should be taught, much more than the existing system of education provided; they were also told that, in terms of practical experience, they were superior to any other body of working men on earth. They might never become scholars, but their natural aptitudes — their instinctive talents for trade, industry and agriculture — made them a special breed of men. Their labour, whether mechanical or manual, made Britain the leader of nations. Should they gain new skills and increase their knowledge, their contribution to society would be undeniable. And how was this to happen? Self-knowledge would enable the working man to do a better job and to choose his craft wisely in the first place. Hence the determinism of phrenology was turned to good account: while some men were destined to be artisans it was also true that all men were 'born with an aversion to idleness'. All men of sound mind, that is. Those who belittled the honest and industrious worker did not understand the 'respectability' of the individual' which, according to Abram Combe (George's brother and an associate of Owen) the 'old system of society' ruthlessly denied.[19] Work was so much a part of human nature that when William Ellis asked 'What Am

I?' his next question was 'What Ought I to do?' By phrenological standards the response was entirely predictable: man must work as skillfully as he can, guided by his knowledge and taking care not to consume the fruits of his labour as soon as he produced them.[20]

Of course the gospel of work did not originate with phrenologists. Nonetheless it owed much to their avid support. Dr Spurzheim himself set the tone for their devotion to this social gospel. When in 1823 he was offered an honorary membership in the Edinburgh Phrenological Scoiety, he replied that although flattered by the idea he did not think it was a good one. In fact he could not approve of honorary memberships of any kind. If a person wished to be associated with a society at all, he must be willing to assume responsibility within it and to work hard for it.[21] Ideally it was the same with society at large: the benefits and joys of membership came only through honest work. Dr Epps and Charles Bray developed the notion still further, by suggesting a reduction of privilege and sinecures which were indefensible in view of the mental traits of the men who held them. It must have seemed to many mechanics that phrenology lent a psychological argument to their vision of a more equalitarian society.

And perhaps the materialism of phrenologists appealed to them as well. The readers of the *Constitution of Man* were assured that the earth provided all the materials to supply human needs and that man could harness all things for useful purposes. That was Nature's plan: man was given all he required; and provided that he cultivated, adapted and restocked the resources which he drew upon, an abundant life was guaranteed. Everything depended, in a most equalitarian way, on everyone doing his fair share of the job. Work was not, as some clergymen maintained, the punishment for sin; nor was it man's everlasting curse. It was instead the moral right and duty of all and in harmony with God's laws.[22] Admittedly there was a large dose of economic liberalism attached to phrenology's praise of work, and it is also true that phrenologists owed much to Adam Smith's concept of government. When George Combe and M. B. Sampson discussed the question of negro slavery, they agreed that society was wise to encourage (if not enforce) that amount of labour which was beneficial to the individual and useful to the community, regardless of colour or race.[23] Because society was the aggregate of working people, the proper role of government was to encourage employment and self-reliance. For this reason Simpson decided that it was not morally right for Henry IV of France to have promised a chicken to every peasant household in his kingdom; if all peasants were first given the opportunity for regular work and good wages, they should then be able to provide for themselves. If only better working conditions and better wages could be secured, it was within man's power to transform his condition from 'one of gloom and misery to one of hope and encouragement';[24] by dint of honest labour and labour's just reward it was possible for artisans to share in the economic well-being of their employers.

The journey from gloom to hope required more than honest labour. Phrenologists were well acquainted with the working conditions of their day; on at least one occasion, in Dundee in 1832, local phrenologists helped workers to draft a petition for factory regulation.[25] Phrenologists were not worried about an unproductive, undeveloped Britain; what worried them was the time man needed to develop his moral faculties. They were concerned about the length of the working day and they were active in the campaign for shorter hours. In the *Lectures on Education* and the *Philosophy of Education*, Combe and Simpson discussed the mental consequences of 'labour carried to excess'. As amateur psychologists they despaired of man's moral improvement after twelve or more hours of hard work.[26] In their view the whole purpose of power looms and steam engines was not to increase the overall wretchedness of the majority, or add to the wealth of the few, but to give more leisure time to all. With the diffusion of manufactured goods and with the abridgement of human toil, every person could expect a 'reasonable leisure' in which to see to the needs of the mind. To the extent that this did not occur, Simpson thought the Industrial Revolution intrinsically evil.[27]

While phrenologists liked to think that they could help operatives in the factory express their grievances, the fact is that they were more effective (and certainly caused a greater sensation) with their ideas on what to do with the leisure time already won. Believing man an intelligent and moral creature who enjoyed a 'relation to a higher and invisible order',[28] phrenologists insisted on a Sabbath freed from the dull routines of industrial society. But this did not mean a day filled with devotional practices. Those too were thought to be dull routines. What the phrenologists wanted was a broader range of activity on the Lord's Day, and generally they did not bother to define that 'higher and invisible order' which was the 'proper object of *veneration*'. In addressing working men's clubs, phrenologists spoke frankly of their dissatisfaction with the way the Sabbath was observed: as a day of indolence when people sat for hours to hear the admonitions they heard all week in another form; a day when they were obliged, by the social custom of the land, to endure 'doctrinal discourses on what have been miscalled the mysteries of religion'.[29] The best use of Sunday was a complete break, refreshing to mind and body, from the usual routine. Obviously Providence (the 'higher order') had intended Sunday for field-trips and exercise. In saying this, phrenologists incurred the clerical wrath, particularly in Scotland where in 1849 they decided to wage battle over the question of passenger services on Sunday trains.

It was certainly not among the more momentous issues of the day, but the train service signified a great deal to local phrenologists. Led by Charles Maclaren, they had always opposed the 'bigots and hypocrites' who tried to 'terrify other people into accepting their sectarian doctrines about the Sabbath'.[30] Phrenologists rejoiced several years earlier, when a Sunday passenger service was introduced; now they were afraid that the suspension of that service represented a stiffening of opposition to the social needs (and to the mobility) of ordinary people. Such were the fears circulated among artisans in the Lowlands by Robert Cox, a leading Scottish phrenologist. As a writer and journalist, Cox (1810–72) welcomed the chance to organise public opinion. In no time he won the reputation as a baiter of clerical gentlemen who, much to his delight, reacted as he expected them to. They viewed his activity as the thin end of the wedge of atheism, and they knew all about his variety of materialism.

It was his sporadic study of anatomy which brought Cox to the new physiology of the brain. Certainly this was a common path to phrenology, particularly at Edinburgh. Yet Cox was no ordinary phrenologist. Twice the editor of the *Phrenological Journal*, he was also the nephew of George Combe. Cox enjoyed every minute of his fuss over the Sabbath. But it was never simply a matter of train schedules. At issue was the influence of churchmen (of all descriptions) in the mental climate of the times. Even after the train service was restored, Cox continued his barrage against the clergy, accusing them of trying to prevent the true mental and moral improvement of the people. The focus on Sunday (which he retained in a series of popular pamphlets) was a clever tactic on his part, for it suggested that people could not really claim one full day to themselves. He implied that Sunday was a kind of mental protectorate which certain churchmen, for sectarian reasons known only to themselves, were trying to force upon the whole of society.[31]

In his campaign Cox was very sensitive to the temper of artisans. He knew what the findings of Henry Mayhew and the census of 1851 were soon to illustrate, that religious apathy was fast becoming a national trait; a more negative sign was the small groups of working men, especially in the larger cities, who openly denounced the habits of organised religion. With various people – the indifferent, the hostile and the secularists – the arguments of Robert Cox found their mark. What they opposed was that the knowledge, the routine and the emotions of Sunday should be so narrowly curtailed. What they wanted (both figuratively and literally) was more room to manoeuvre, physically and mentally. The mobility of the spirit and the circulation of ideas were just as much at stake as the physical conveyance of people, and real social freedom required the recognition of widely differing emotions and sentiments on Sunday. Put in such psychological terms, Cox's argument sounded convincing. His approach was similar to that used by all phrenologists with respect to the merits of self-knowledge, work and temperance: these social gospels were now given the irrevocable blessing of science. What J. L. Levison, Henry Turner and Robert Cox accomplished before their listeners in the institutes and assembly rooms was simply this: they helped to substitute a decidedly secular and apparently scientific morality for the older evangelical sort. It was an intellectual process of changing the foundations while the edifice of social opinions and values remained intact.

The net result of underlining what people already believed or wanted to believe was to produce a rather soothing effect. In fact one can almost say that phrenology's support of Victorian social gospels helped to create a certain smugness of mind – even among the artisan class. Their reflections could lead only in one direction. Was it not true that their country pursued self-improvement and self-knowledge more fervently than any other? Other people, it was true, recognised the merits of temperance, but where was this cause taken most seriously? As for the gospel of work, of course no reasonable person anywhere imagined that life was a garden of Eden; but whose national genius now dominated the world of industry and invention? The answer was enough to excite *self-esteem*. The attainments of the British mind were not easily duplicated elsewhere. As far as the reading and listening public in Britain was concerned, the science of Gall and Spurzheim carried the notion of race a long way.

Phrenologists appreciated the long duration of human development and it was obvious to them that human progress was nowhere the same. It depended above all on the cerebral structure of different tribes. Most leading phrenologists had large collections of skulls, animal and human, of real bone and of plaster. They liked to emphasise that their science was a comparative one: it allowed the contrast of individual faculties with each other and of the whole skull of one race with that of another. The collections included specimens from all over the world, and the most recent encounters with the aboriginal peoples of the south-west Pacific proved to phrenologists that there was a definite correlation between the shape of the head and the degree of civilisation.[32] The shape of the head also accounted for the slight peculiarities which one noticed among people of the same general tribe. The Scottish and Irish brains had special traits which set them apart from the English, while all British minds were distinguishable from the Mediterranean and the Asian. The old national stereotypes which Englishmen enjoyed making of their continental neighbours were therefore absorbed into phrenology and given a deeper meaning. The French were vain because they were Celts, and Celts usually had large faculties of *self-esteem* and *love of approbation*. There was also a relative lack of *benevolence* in the French brain, which suggested that France could always be counted upon to produce soldiers to instigate wars abroad and rebellions at home. Under the circumstances it was difficult to deal with the French, however much one might enjoy their artistry; George Combe thought that the only hope of France lay with its non-Celtic citizens of the north and east who, being more teutonic, had visibly better moral and intellectual organs.[33] The Germans themselves, while 'profound thinkers', were not a practical race: they were filled with serious emotions which could make them unduly fierce.[34] Closer to home, it was evident that not all of Ireland was overcome by crime, want and superstition (the last caused by a large faculty of *veneration*) because at least the people in the north of the country were intelligent, industrious and moral.[35] In fact, as British phrenologists explored the world of heads, they discovered how virtuous were their own by comparison. It was the prediction of Robert Verity, a well-to-do phrenologist and physician, that the destiny of the British race was to preside indefinitely over the world of science and letters.[36]

The proof which phrenology lent to popular notions of race and nationality tended to make phrenologists little Englanders. The wisest policy, they thought, was one which insured a happy cooperation among the mentally advanced peoples of the Anglo-Saxon world and western Europe. As long as peace obtained in this quarter there was nothing to fear. Not even the Czar, for all his supposed ambition, threatened British interests, for he still had to contend with the mental deficiencies of his own people. 'Russia can triumph over Britain only by becoming superior in intelligence', concluded one writer who also discounted the possibility.[37] It was the logic of phrenology that the greatest peril rested with the most primitive races. The danger was a subtle one. It involved the long-term effects on British character of intermarriage with inferior peoples. The resulting strain, no matter how few in number, would undoubtably hinder the mental progress of Britain itself should any of them choose to settle there. That was the disadvantage of a vast tropical empire. Granted, it had been an easy empire to secure because 'a handful of Europeans overcomes in combat, and holds in permanent subjection, thousands — nay, millions of people so wanting in force of character.'[38] The culture of the Hindus, for example, was a 'stationary one' because of their deficient moral faculties. The *Phrenological Journal* was sweeping in its judgment of all those now called Afro-Asians: 'The people of Asia early arrived at a point comparatively low in the scale of improvement, which they will never pass . . . and the history of Africa, so far as Africa can be said to have a history, presents a similar phenomena. The annals of the races who have inhabited that continent, with few exceptions, exhibit one unbroken scene of moral and intellectual desolation.'[39]

The idea that whole continents of people could never catch up with the intellect of the Anglo-Saxons is but one aspect of the 'fatalism' of phrenology. True, man was a rational creature, but (in Orwellian terms) some men were more rational than others. In the *Philosophy of Necessity* Charles Bray explained that although man did as he pleased, his pleasures and his conduct were largely predetermined by his mental constitution and by the circumstances in which he found himself. Bray did not attempt to measure the relative weights of these two forces, the one inherent and hereditary, and the other external and accidental; but as a phrenologist he knew that man's environment must be the weaker influence. It was a vulnerable position to take and the phrenologists could not easily avoid it. Ultimately they had to fall back on their understanding of Nature. The human species was one of many which Nature had set in operation; if all the species were subject to struggle and death, and if over the centuries the brains of some men developed more fully than others, was it not proper for the superior to supplant the inferior in all cases? History had shown that in racial terms this meant the subjugation of the black and red peoples to white, and the feebler whites to stronger members of their own race. This is what Watson (although politically a democrat) described as the 'natural course of events'.[40] It explained the conquest and the decline of the American tribes and the wretched condition in which Captain Cook found the natives of Australia. Indeed, the definition of race was perhaps the most durable which phrenology provided. Writing at the end of the century, Alfred Russel Wallace, one of the principal theorists of evolution, thought that phrenology accounted for all the national and racial traits which neither the physiologists nor the anthropologists ever understood.[41]

Given the cerebral determinism of phrenology, there were geographical limits to the value of 'Self Help' and other British social gospels. For the Europeans and particularly for the Anglo-Saxons, Self Help was a credible alternative because people were able in time to restrain or redirect the powers of their minds. The Africans and Asians were not considered likely to proceed beyond their present attainments, and the Arabs were hampered by the propensity to steal, cheat and destroy.[42] But while phrenologists doubted that the dictums of Dr Smiles would ever be of much use to Africans, Asians or even to Mediterranean peoples, the racialism of phrenology was not altogether vicious. Phrenologists denounced negro slavery as the 'offspring of the basest propensities' and they saw no reason for their country to bear the white man's burden.[43] Theirs was another way of saying that imperial policy should involve colonies of British settlement rather than of large native populations, and they decided that any addition to the Empire was worthless if it were already inhabited by a different race. In their correspondence Combe and Cobden agreed that India was 'an immoral appendage of Great Britain' which injured the mother country more than it enriched her.[44] This was not so much the racialism of hatred and contempt as it was the racialism of self-superiority and aloofness.

The racialism of phrenology was largely the result of attempting to understand the practical limits of man's progress, self-knowledge and self-improvement. In giving racial stereotypes a new foundation, phrenology was making full use of ageless and popular notions whose utility was universal. They were common to poets and teachers, lawyers and reformers, artisans and parsons, phrenologists and antiphrenologists. Racial expressions entered the language of almost any argument. When Stephen Seedair attacked phrenology and upheld Christianity as the 'only true constitution of man', he praised the superior endowments of 'the Norman-Saxon mind'.[45] The British readers of Seedair's pamphlet and of Combe's *Constitution of Man* accepted praise from any quarter; and it was the season for compliments. The race which had vanquished tyrants from the continent, become the first industrial power in the world, and encouraged the education of the artisan at home was already convinced of its destiny. Phrenology was attractive because it explained why the British were so fortunate a people and what they must do to remain so. And yet the paradox is that phrenology never taught a complete equality of fortune even for the British. There was a sexual selectivity as well as a racial one. Why then should so many women have shown greater enthusiasm for the science of Gall and Spurzheim than their husbands did?

There was, after all, 'less vigor in the female intellect'.[46] Phrenologists found it easy to justify man's superiority in Victorian society. They thought that decision-making should always rest with the man, because the female reflective powers were small by comparison. 'Women do not extend their reasonings beyond the range of the visible world', noted the *Phrenological Journal*. 'Nor do they make any great or daring excursions into the regions of fancy.'[47] Still, the ladies came. Often they comprised well over half the audience. On his Irish tour in 1816 Spurzheim encountered large female audiences drawn from every rank of society, but he was not surprised. He felt that the ladies were 'turning their minds to scientific pursuits' while the gentlemen tended to business or physicians went to suppers at their clubs.[48]

True, the pursuit of science (in its most general and entertaining form) had something to do with it. Sometimes a series of lectures by a visiting phrenologist was the only opportunity the ladies had for instruction in physiology or the natural sciences. And as phrenologists often philosophised about education and other social issues, it is hardly surprising that the lectures should have been so well attended. James Simpson observed that 'many ladies of middle rank seeking useful knowledge' came to his lectures, which they supported as a happy alternative to an expensive press.[49] Women were also aware of the improved education of the phrenologist schools: institutions which took into account the needs and talents of individual boys and girls instead of giving all children an equally bad education. The emphasis which William Ellis and his wife placed on the 'moral trainings and female accomplishments' was only one facet of the specialised and personal treatment which young ladies could now obtain. In their pamphlet 'Progressive Education for Young Ladies', the Ellises surmised that women 'must do what they love to do, and not by rote'. Those who understood the special values of the female mind (especially that of *philoprogenitiveness*) realised that women also required a certain amount of practical, scientific knowledge.[50]

If 'useful and entertaining knowledge' was the objective, it is nonetheless true that the accent was often on 'entertaining'. Certainly women of all ranks were (like the menfolk) attracted to phrenology and later to phreno-mesmerism because the lectures, complete with graphs and portraits of famous persons, represented a pleasant and unusual *divertissement*. Where there were few opportunities for regular entertainment, as in parts of Ireland and Scotland, and where social gatherings were usually religious affairs, women attended phrenology lectures in particularly large numbers. Among the more serious, phrenology served as a stepping stone to various social crusades. The lectures which Andrew Combe gave on the cerebral effects of hard drinking were supposed to have added to the female membership of more than one temperance society. It was important for women, whose education was almost entirely domestic in character, to learn something of human nature before they were able to support such causes as temperance or slavery abolition, and phrenology lectures were frequently the quickest way of obtaining such information. The 'ladies of middle rank' who attended Simpson's lectures sought out phrenologists as the new prophets of man, who bore Nature's commandments not in tablets of stone but in maps of the skull.

The most practical information dispensed by phrenologists was that which pertained to health and living conditions. Phrenologists were concerned with the total well-being of man and not merely his mental activity; their study of physiology and anatomy made them spokesmen for the social (if not political) emancipation of women. Phrenologists in Britain as well as the United States and Scandinavia wanted to bring women out of their cosy domestic confinement: to allow them to join and lead public associations, to provide opportunities for their exercise and education, and to liberate them from the dictates of fashion. Dr Andrew Combe's *Physiology*, together with numerous articles in the *Phrenological Journal*, all decried the stylish expedients which restrained women and which violated the laws of health. In America Dr Caldwell called upon women to discontinue the 'irrational and unnatural use' of corsets; as a physiologist he realised that corsets prevented the free passage of blood and promoted 'many forms of painful and annoying disease'.[51] For those women who shared Caldwell's opinion, the laws of health and the study of physiology, which were always central to the phrenological message, held the promise of social freedom. Women were as equally disposed as men to place their faith in a science which enabled them to understand their proper relationship to Nature, and they may well have appreciated the fact that the physiologists who championed female liberties in the world of fashion were often phrenologists into the bargain.

What was true of the ladies was, in many respects, true of all those who wanted to believe in phrenology. They approached the science as they might have approached a buffet table, picking and choosing what they desired or what seemed palatable to common sense. Perhaps this is why 'Candidus', that anonymous subscriber to lectures in Edinburgh, refrained from condemning what he was unable to disprove, and why he praised instead the practical knowledge which phrenologists bestowed. Was it craniology or simply good sense to advise householders to examine the 'dispositions and temperaments' of prospective servants.?[52] Was it phrenology's head-reading or its pragmatism which satisfied members of the Central Society for Education when they read works by phrenologists?[53] The reader of the *Phrenological Journal* could possibly ignore the premises of any article, if he thought them too weak; he could subtract many tenets of the phrenological credo. But somehow the conclusions remained plausible and (in spite of the methods) remarkably modest. Like Oliver Wendell Holmes in America, many people thought that phrenologists were simply preaching 'good sense under the disguise of an equivocal system'. It was the smallest concession anti-phrenologists could make, and they made it easily enough. Had they been more honest they might have admitted that the system of phrenology was not so outrageous. They could (and many of them did) follow Coleridge's example, reserving judgment until 'some proper names are discovered for the organs'. This would allow them, as it did Coleridge, to say that 'all the coincidences' detected by phrenology 'could scarcely be by accident'.[54] As long as men and women of every class and of every profession were ready to concede that much, phrenology would survive as an explanation of human conduct.

There were many causes for the wide if intermittent popularity of phrenology in the nineteenth century. It was an easy philosophy, expressed in ordinary language. It was a guide to reform and to knowledge; it was a new basis of morality. It was logical and slightly mysterious, precise but flexible, awesome in judgment and yet humanely hopeful. It meant amusement and improvement, common sense and social liberation. The merits were many but they were fleeting. The next stage in the development of psychology, far from vindicating Spurzheim and Combe, erased almost completely the support they once enjoyed in scientific circles. The research of the 1860s and 1870s (and particularly the work of Alexander Bain) had the effect of burying most of the moral and philosophical pretensions of Combe's 'great system'. Of course phrenology was not eliminated from the social scene: it entered the twentieth century as little more than head-reading, a pleasant (and quite respectable) distraction at resorts like Brighton. The curiosity is that long before this happened the phrenologists still blamed an 'adverse press', and not the more powerful ideas of Professor Bain, for their ill fortune.

Notes

1. Watson to Combe, 3 May 1839. NLS 7252/147.
2. 'Candidus' to Combe, 5 July 1836. NLS 7239/106.
3. John Chambers, of the *Chambers Journal*, quoted on the title page of the *American Phrenological Journal*, no.14 (1857).
4. Henry Clark, *Christian Phrenology; or the Teachings of the New Testament respecting the animal, moral and intellectual Nature of Man* ... (Dundee, 1835), p.16.
5. J. T. Smith, *Synopsis of Phrenology: directed chiefly to the Exhibition of the Utility of the Science to the Advancement of Social Happiness* (London, n.d.).
6. J. L. Levison, *Mental Culture; or the Means of Developing the Human Faculties* (London, 1833), pp.12–13 and 210.
7. Dr John Abercrombie, *Culture and Discipline of the Mind* (6th ed.: Edinburgh, 1837), p.25.
8. Henry Turner, *Phrenology: its Evidences and Inferences, with Criticisms upon Mr Grant's recent Lectures* (2nd ed.: Sheffield, 1858), p.3.
9. [John Epps] *Diary*, ed. E. Epps (London [1875]), p.131.
10. Combe to Frances Kemble, 23 July 1830. NLS 7384/520.
11. *Journal*, VIII, no.40 (1834), p.606.
12. Simpson, *Philosophy of Education*, p.10.
13. J. L. Levison, *Lecture on the Hereditary Tendency of Drunkenness* (London, 1839), p.50.
14. *ibid*.
15. *Journal*, VI, no.21 (1829), p.45.
16. *ibid.*; also Andrew Combe, *Observations on Mental Derangement* ... (Edinburgh, 1830), *passim*.
17. William Ellis, *Education as a Means of preventing Destitution; with Exemplifications from the Teaching of the Conditions of Wellbeing* ... (London, 1851), p.9.
18. Combe, *Relations between Science and Religion*, p.181.
19. Abram Combe, *An Address to the Conductors of the Periodical Press* ... (Edinburgh, 1823), p.27.
20. [William Ellis] *What Am I?* (London, 1852), p.36.
21. Spurzheim to Combe, 14 April 1823. NLS 7211/85.
22. Combe, *Relation between Science and Religion*, p.193.
23. Combe to M. B. Sampson, 2 December 1844. NLS 7388/810.
24. Combe, *Relation between Science and Religion*, p.193.
25. George to Andrew Combe, 19 July 1832. NLS 7385
26. Combe, *Lectures on Popular Education*, p.63.

27. Simpson, *Philosophy of Education*, p.43.
28. *Journal*, VI, no.26 (1830), p.555.
29. Simpson, *The Necessity of Popular Education as a National Object*... (Edinburgh, 1834), p.31.
30. Maclaren to Combe, 23 February 1842. NLS 7265/36.
31. Among Cox's published works were his 'Plea for Sunday Trains' (Edinburgh, n.d.), the *Sabbath Laws and Sabbath Duties, considered in relation to their natural and scriptural Grounds* (Edinburgh, 1853) and the *Whole Doctrine of Calvin about the Sabbath* (Edinburgh, 1860).
32. When in America George Combe sold his collection of skulls and casts to a member of his audience in New Haven, Connecticut, who in turn presented it to the Medical Department of Yale College in 1841. Dr Gall's immense collection was sold to the Museum of Man in Paris. The Nazis (and particularly Himmler) were interested in this collection, and their argument was the usual one: Gall was German and therefore his collection belonged in Germany. Parts of the collection, however, were either lost or destroyed in the transfer.
33. Combe to Cobden, 25 January 1847. B.M. 43,660/68 (C.P. XIV).
34. Combe, *Notes on the New Reformation in Germany*, p.1.
35. 'Cursory Remarks on Ireland', *Journal*, II, no.6. (1825), pp.161–6 and 176.
36. Robert Verity, *Changes produced in the nervous System by Civilization* (London, [1839]).
37. *Journal*, XIII, no.65. (1840), pp.375–6.
38. *ibid.*, II, no.5 (1824), p.7.
39. *ibid.*, p.3.
40. Watson to Combe, 14 May 1847. NLS 7288/127.
41. Wallace, *The Wonderful Century: its Successes and its Failures*, (London, 1898), p.189.
42. Dr Charles Caldwell, *Thoughts on Physical Education, and the true Mode of improving the Condition of Man*... ed. Robert Cox (Edinburgh, 1836), p.17.
43. *Journal*, VIII, no.34 (1832), p.80.
44. Combe to Cobden, 13 July 1848. B.M. 43,660/122 (C.P. XIV).
45. Stephen Seedair, *A Tract for all Time. The Christian or true Constitution of Man, versus the pernicious Fallacies of Mr Combe and other materialistic Writers* (Edinburgh, 1856), p.21.
46. *Journal*, II, no.6 (1825), p.278.
47. *ibid*.
48. Quoted by Watson in *Statistics of Phrenology*, pp.120–1.
49. Simpson, *Philosophy of Education*, p.245.
50. 'Progressive Education for Young Ladies', a seven page pamphlet in the Combe Papers, NLS 7279/109–112.
51. Caldwell, *Thoughts on Physical Education*, p.115.
52. *Journal*, VI, no.22 (1829), pp.213–4.
53. Watson to Combe, 10 September 1839. NLS 7252/151.
54. *Table Talk and Omniana of Coleridge*, 24 June 1827.

Chapter V Transmission and Schism

In 1823, in the wake of fresh attacks on phrenology in Scotland, Dr Spurzheim wrote a letter on tactics to George Combe. Spurzheim advised that phrenologists must always be ready to meet the rough and abusive language of their opponents calmly, avoiding any display of temper and referring constantly to their observations of Nature. Spruzheim had no doubt that this policy, if followed patiently, would have a 'good effect on honest minds' and confound all those who waited upon phrenology's shameful retreat.[1] For those convinced of the truthfulness of the new science, it was the most reasonable course of action. At the same time, phrenologists knew that calm answers would never be enough. In the long term, the only way to win the battle was to gather support throughout the country. No phrenologist thought his science was meant for an elite group of philosophers and academicians; it was a science for all people, and the argument of sheer numbers was the best means of defence. And how was this to come about? Did it mean that each phrenologist had a weekly quota of converts? Should they resort to the tactics of the Bible Society, as Dr Elliotson suggested? Anti-phrenologists might come and go, causing whatever trouble they could; but how was phrenology to find its friends? The only way was through time-tested methods of propaganda: by forming local societies, by printing journals, and by conducting lecture tours. Properly used, these techniques would gain phrenology the allegiance of 'honest minds' everywhere, and phrenologists themselves, riding the crest of popular acclaim, would then feel no need to return any answer to their critics.

1

The paradox of the local societies is that, while they were fairly numerous, they were not particularly important to the spread of phrenology. At the beginning, however, the new philosophy needed as many spokesmen as possible, and it was hoped that the societies might become seminaries for the selection and training of zealous missionaries. Spurzheim and Combe realised that the public must learn phrenology from articulate and orthodox phrenologists, and not secondhand through the attacks of its unprincipled opponents. Only the best men should be sent forth to gather the harvest and it was necessary to choose disciples wisely. 'The choice of members is of utmost importance', Spurzheim cautioned. 'We must remember that Christ was betrayed by an ill-chosen subject.'[2] The reception of new members was indeed a careful business. All applicants to the phrenological societies had to bear recommendations from respected members, and the novices had to be acquainted with the general principles of the science as a preliminary to full membership.[3] In addition, the candidate had to submit his head to a thorough examination – a precaution which Dr John Epps, who joined the Edinburgh society while he studied medicine, thought only reasonable.[4] The cost of membership may not have been as agreeable as the induction process. Subscriptions were generally high. The Scottish societies charged an annual fee of at least thirty shillings as well as an entrance (admission) fee, and any member who neglected his financial obligation for two years was liable to lose his membership and was obliged to start all over again. There were more flexible provisions for 'corresponding members' who lived at some distance from the local chapter and accordingly paid a smaller subscription than the 'ordinary members'. The fiscal expectations of membership (if not the inspection of heads) helped to insure quite respectable societies, including, as the *Phrenological Journal* pointed out with reference to Edinburgh, 'many professional and scientific gentlemen, and several eminent artists'.[5]

The local societies were almost entirely on their own. Simpson once likened them to isolated outposts at the mercy of the hostile wilderness. The inquiries which Hewett Watson undertook for the *Statistics of Phrenology* indicated that there were societies in half the counties and in cities and towns of all sizes, ranging from 'very insignificant places' like Bakewell and Dingwall, to Stirling and Southampton, Hull and Belfast, Glammis and York.[6] Many of the societies had their own libraries and museums complete with casts or skulls, and almost all boasted at least one member from the medical profession. What the societies really needed, however, was a permanent national organisation to bind them together. Without a superior national body, the presidents of the local societies acted rather like powerful bishops, for there was no one to call them to order. The societies were loosely related by their general profession of the phrenological creed, by their reverence for the truths revealed in the *Constitution of Man*, and by their esteem for Spurzheim and Combe. It was virtually impossible for the president of one society to direct the membership of another unless it had no leader of its own, and for all the respect which he enjoyed, the most that George Combe could ever do was to caution or encourage, cajole or suggest, always carefully and politely. Each local society was free to determine its position, with respect to both its enemies and to its uncommitted spectators; and in the world of phrenologists, each society was also free to decide whether it preferred the style of Gall and Elliotson to that of Spurzheim and Combe, union with London or with Edinburgh.

While the local societies were not subject to any external supervision, their activities were almost everywhere the same. If the members met regularly, they might read essays on the phrenological features of Shakespeare's characters or discuss the mental faculties of the preschool child. Inevitably there were cerebral descriptions of famous criminals and the occasional visit by a phrenologist of national repute, such as Watson, Simpson or the Combes.[7] Unfortunately the little information which we have about the societies suggests that they had no real sense of mission or *esprit de corps*. They tended to think of their organisations as clubs rather than as parishes of a great faith, as places of leisure and social assembly rather than as the local headquarters of a national crusade. It is true that the official purpose of their meetings was to inspire the faithful and to convince the newcomer of phrenology's usefulness, but this intention does not appear to have been carried out regularly. When they met, the societies often included on their agendas a number of non-phrenological topics. There were many reasons for this: for one thing, phrenologists were sometimes members of scientific and literary clubs, and the business of one group frequently carried into another. Often too, the full membership of a phrenological community was divided along class or professional lines and when they gathered for their evening meetings, members took the opportunity to talk about problems they had in common. Watson noted without comment in the *Statistics* that the Phrenological Society of Glasgow was established in 1829 and that a separate society for Glasgow operatives was formed later.[8]

A more critical weakness was the fluidity of membership. Subscribers came and went, and George Combe resigned himself to receiving countless letters from part-time phrenologists who confessed that they no longer attended meetings or did anything on behalf of the science. In very few of the societies did the rank and file members take an active role. They were thankful for the aid which phrenology gave towards the making of a general philosophy, but they did not care to raise either their phrenology — or the study of philosophy — above the amateur plane. Besides, other societies and associations satisfied the same social and scientific needs at a lower cost. In Edinburgh there were probably more phrenologists (of varying degrees of devotion) in the Association for Providing Instruction in Useful and Entertaining Knowledge than in the Phrenological Society itself. As their enthusiasm declined, phrenologists transferred their attentions and merged with other groups, as the Belfast phrenologists proposed to do with the Natural History Society in their city.[9] By 1850 the societies were practically moribund and no new organisations were being established. A few of the local societies survived in skeletal form through the fifties, the only remaining agents of a pseudo-science whose devotees in the 1820s believed might become the national philosophy.

A hostile press stood in the way. When the papers did not ignore phrenology, they ridiculed it. *The Times* in London and the Edinburgh *Review* were the most savage opponents; they never failed to lampoon the 'crazy craniologists'. In Scotland the battle was particularly brutal, thanks to the personal emnity which existed between George Combe and Francis Jeffrey, editor of the Edinburgh *Review*. By 1830 their fierce exchanges gave way to a polite cold war, because the new editor of the *Review*, McEvey Napier, decided to ignore phrenologists rather than tease them. Elsewhere, the coming and going of phrenologists was reported briefly but fairly. Hewett Watson estimated that in the whole of Britain there were twenty-five to thirty newspapers which regularly carried reports of phrenology lectures or meetings.[10] Among these (at least for a while) were the Birmingham *Times*, the Wolverhampton *Chronicle*, the Lincoln *Gazette*, and many smaller papers such as the Chatham *Telegraph*. Only one major newspaper, the *Scotsman*, edited by the phrenologist Charles Maclaren, endorsed phrenology without reservation. It was a bleak picture, and as the momentum of phrenology began to fade, George Combe blamed the national press, which he thought 'ignorant, dogmatic and in many cases dishonest and trampling down the truth'.[11]

There was only one way around this obstacle. If the opposition of leading British papers proved so unrelenting, and if phrenology was to enjoy favourable and regular attention in the press, phrenologists would have to publish it themselves. It was to meet this need that William Scott, James Simpson, Dr Richard Poole and George Combe gathered their resources in 1823 to start the *Phrenological Journal*. The objectives were given in an introductory statement which indicates that they were fully aware of the challenges facing their science. The *Journal* promised to present orthodox phrenology clearly and without compromise and to defend the 'sound morality of pure religion' – an indication that trouble was expected from more traditional-minded clerics. The *Journal* also pledged to bring speedy justice to those 'pseudo-phrenological writers who attempt to pervert the science by a contrary course' – a declaration of war against those who adopted phrenology for mere monetary profit. Lastly, in keeping with the full title of the *Phrenological Journal and Miscellany*, the founders promised to offer 'lighter material' suitable to the tastes of an enlightened reading public.[12]

While the *Phrenological Journal* began confidently, it shared with the local societies a diminishing popularity. Far from matching the hopes of its editors, the *Journal* never achieved a national reputation and, as Combe intimated to J. L. Levison in 1829, the number of subscriptions did not cover the costs of printing. Early subscribers were attracted by the 'sheer love of novelty and the expectation of wonders' but fell away weary of the *Journal's* repetition and failure to 'apply phrenology'.[13] Watson, always critical of his friends, later complained that the *Journal* simply was not lively enough; it carried too many 'heavy articles'.[14] And he probably had a point. If people were drawn to phrenology by their distrust or their disbelief of traditional philosophy and metaphysics, was it really wise for the *Journal* to feature a thirty-seven page article on 'The Comparative Merits of the Mental Philosophy of the School of Reid and Stewart'?[15] For better or for worse, the editors always felt obliged to prove the claims of phrenology as a viable philosophy, and too frequently this involved lengthy comparisons which may well have tested the patience of many readers.

By the 1830s the editorial board of the *Journal* was meeting the costs of production very largely out of its own collective pocket. They continued to bear the expense for more than ten years, always fearing that the end of the *Journal* would represent an irreversable defeat for phrenology in Britain. In the last analysis, the taunts of the *Edinburgh Review* and *The Times* did not bother the subscribers or the occasional readers of the *Phrenological Journal*. Of much greater embarrassment was the damage caused by fellow phrenologists. In 1843 the society in London, led by the eccentric Dr Elliotson, established a journal of their own called *The Zoist*. The problem of this journal was its dedication to other fringe sciences, notably mesmerism; and it was not long before the two papers, as representatives of the two main branches of phrenology, became rivals for the same limited audience. There were bitter exchanges, and the feud absorbed considerable energy on both sides. The quarrel had two immediate effects: financially it prevented either paper from gaining support as the national voice of phrenology; and the *Phrenological Journal* was the first to go under. More importantly, the rivalry of the two journals meant the formal division of phrenology and created confusion among those expecting to find a science built upon indisputable principles of common sense.

George and Andrew Combe, who had each been editor in turn, saw the problem rather simply as one of maintaining the *Journal's* circulation at any cost. Where was the money to come from? Combe thought there were two alternatives. The first was to ask for annual contributions 'from those who really wish to support the cause', but he and Levison discounted this possibility because dedicated phrenologists were already paying large sums towards membership in the local societies. The second alternative was to urge phrenologists everywhere 'to push subscriptions and at least double the present sales'.[16] But the Combes had repeatedly asked their correspondents to do this, and without success. As early as 1829 the circulation of the *Journal* was so low that Combe estimated the sale of 500 copies would probably suffice to meet the emergency. The only other possibility, aside from asking others to make financial sacrifices, was to change the character of the *Journal* itself.

This idea occurred to Combe and his friends by 1829. In practical terms it meant that the *Journal* should cease to imitate the scholarly style and format of its principal antagonist, the Edinburgh *Review*, and should instead discuss popular subjects in a more popular language. George Combe, for one, was unwilling to do this and he had his way. Because he was opposed to 'diluting phrenology', no substantial changes were made and the *Journal* drifted on, supported by a very small proportion of the royalties from the *Constitution of Man*, until early in 1837 when Combe's friends again raised the question. J. Toulmin Smith, soon to leave on a phrenology lecture tour of America, wondered if the *Journal* was approaching its task in the right way; he thought it was possible for a journal to be too philosophical. Smith's advice was that the *Journal* should become more topical, and he reminded Combe that there was 'no subject upon which phrenology may not be brought to bear'.[17] It was almost an irreverent suggestion to set before the author of the *Constitution of Man*. Hewett Watson was even more explicit. In the *Statistics of Phrenology* he laid the blame for the *Journal's* decline at the feet of the proprietors. Their mistake was that they had never caught the imagination of the public;

they were too elitist in their outlook, taking care to demonstrate their own 'feelings and ideas' while neglecting those of the wider audience of phrenology.[18]

Never before had Watson been so critical and democratic-minded — even with his friends. Naturally Combe was hurt by his remarks, which were made, after all, in the only factual survey of phrenology in Britain. Combe's response was calculated: he offered his position of authority to Watson. At first Watson refused; he knew that, whatever he did with the *Phrenological Journal*, there was no guarantee against financial ruin — which he was far less equipped to meet than were the Combes. But it was one of those occasions when reluctance, however sincere, was the weakest possible excuse. Watson had found fault with the *Journal* and he had announced this finding to the public; he was boxed in by the confidence of his own opinions. It was now up to him to show how the *Journal* might survive, and when he finally agreed in 1837 (after months of pressure from Combe and others), the 'new series' of the *Phrenological Journal* took on a different style.

For one thing, Watson was more aggressive. He never imagined that he could match Combe's reputation as a moralist, and he did not believe in preaching ethics. For one whose training and whose profession was in the science of botany, Watson's ambition was distinctly enormous. He cherished the hope of becoming a great orator, one who transcended Whig and Tory politics and who spoke common sense on all national issues. He loved debating and argument, and now that he was the editor of the *Phrenological Journal* he struck out at critics in every corner. It was time, he declared, that anti-phrenologists become the object of scorn and laughter. 'We are ruder now,' he proudly wrote to Combe.[19] The shortcomings of this policy soon became as apparent to Watson as they had to previous editors. By 1840 it was too late to argue. Phrenology had not acquired the fresh evidence which Gall and Spurzheim predicted and which George Combe still expected. Nor had the anti-phrenologists succeeded in proving all their objections. The debate had long ago reached a stalemate and most of the old contenders preferred instead to await history's verdict.

In the meantime, there was at least one useful change in the format of the *Phrenological Journal*. It lost much of its didactic tone and became the clearing house for all phrenological activity in the country. Under Watson's direction the *Journal* set aside a great deal of space — often as much as six pages — for 'Intelligence', or brief reports from societies in Britain, on the continent, and in America. The *Journal* also reported (and probably exaggerated) the reception given to touring phrenologists, and described George Combe's American odyssey as the beginning of the phrenologists' millenium. Unfortunately, the *Journal's* new force was almost entirely the result of Watson's personal energy. For a few years he dominated the *Journal* much as George Combe had done in the twenties; he even transferred it to London, where he thought the atmosphere was more inspiring than sedate Edinburgh. His style of propaganda was, it is true, more readable and factual — not unlike the style of the *Statistics* which he published in 1836 — and at times it could also be imaginative and impulsive. Watson was always like this: before becoming editor, he had suggested approaching every member of both houses of parliament with the 'evidence of phrenology' in the hope that members might be shown 'how completely attempts at refutation have failed'.[20] But as editor Watson soon tired. He realised that the problem was one of selling phrenology itself rather than newspapers, and this was not the sort of battle he cared to continue. And for all of his efforts and imagination, the *Journal* fared little better than before.

The new style of the *Journal* at least showed a willingness to confront some of phrenology's oldest problems. For almost twenty years phrenologists had complained of the difficulty in convincing people that the science was not a simple matter of 'bumps on the head', and one of the original reasons for the *Journal* was to show how phrenology embraced intellectual and moral philosophy. Watson stretched the horizon of the *Journal* still further, for he thought that only 'partisan papers' focused their attention on a single cause.[21] This was to learn from the experience of the lecture circuit. A 'syllabus of four lectures' given by Sidney Smith in 1838 illustrates how wise phrenologists expanded their topics to meet popular requests. The first lecture in Smith's series dealt with 'morality' in all phases of national life, including the press and politics. The second considered the 'morality of religion, the constitution of the labouring population, the causes of drunkenness, and the use of churches'. The last two lectures treated 'Improvements in Agriculture and Machinery', comparative Christianity, and the role of women in modern society.[22] With the exception of the agricultural topic, there was sufficient room in the series to talk about mental aptitudes and mental progress, which is precisely what Sidney Smith did. Combe might have criticised this approach for 'bringing in phrenology by the back door', but an able lecturer like Smith made the best of it. Watson, while editor of the *Journal*, attempted to do the same thing in black and white.

The popularity of Sidney Smith and other phrenology lecturers serves to point out the greater impact of this form of propaganda over the press and the local societies. The public lecture was invariably the easiest and the most sociable path to phrenology. In Edinburgh George Combe and other phrenologists were often invited to address the 'Association for Providing Instruction in Useful and Entertaining Science'. The name of the group is truly revealing. Phrenologists had every intention of linking phrenology to 'popular and practical science', for they knew that their science did not suffer from being integrated with others. The receipts of the Association at the end of the first year tell the story:

geology lectures	£73
physiology lectures	£89
chemistry lectures	£100
phrenology lectures	£115[23]

According to James Simpson, who also lectured at the Association, the admission fee of 6d. was charged for each lecture in all the series. The only exception was astronomy, which the Association recognised as a reliable money-maker and therefore charged an admission of one shilling. In such places where the premium was on science, phrenology in the twenties and thirties was clearly in demand. It was as entertaining as any other science, and an hour's talk by a reputable phrenologist made a profitable evening for promoters and audience alike. The historian of the Literary and Philosophical Society in Newcastle-upon-Tyne commented that George Combe's lectures there in 1835 'were so great a success that they scarcely cost the Society anything. People were impatient to know what stories their skulls told about them.'[24] The certainty of profit was such (at least in the twenties and thirties) that phrenologists were often invited to perform for charity. The Committee for the Relief of Distressed Operatives knew of no better way to secure funds than to ask George Combe to give a benefit lecture in 1828. The lecture was attended by 600 persons even though it lasted for three hours. Phrenologists were usually agreeable to giving 'charity lectures', for while their journals and societies languished they probably felt that crowded assembly halls meant that phrenology was reaching the public after all.[25]

Everyone, it seems, was more enthusiastic about the lecture as a means of propaganda and entertainment. Attending the occasional lecture involved less effort than regular membership of a club or society and it was more convivial than reading journals or papers. Throughout

Britain the number and variety of subscription lectures multiplied, and those given by phrenologists were part of a fashionable pattern. The topics were generally scientific in character; a few were notably philosophical. Early in 1815 the *Morning Chronicle* drew the attention of its readers to a course of lectures 'on the passions and affections of the mind, their connection with the organisation of the body and their influence on savage and civilised life'.[26] It was an ambitious course, but in effect the phrenologists were doing the same thing. In sociological terms there was one important difference. Phrenologists actively sought — and obtained — audiences more representative of society as a whole. Those who attended the lectures sponsored by the Russell Institution or the Surrey Institution in London, for example, were usually from the upper middle class; and they probably had to be, for the lectures on electricity and electro-chemistry at the Russell Institution cost £1 4s. for regular subscribers.[27] The skilled artisans, school teachers and businessmen who heard Combe and Simpson at the Association for Providing Useful Instruction in Edinburgh rarely paid more than 9d. for each lecture. The twenties and thirties were, for the most part, years of feasting for professional phrenologists, and the risks and expenses of journalism were almost forgotten in the profit of lecture tickets.

This is where the money was to be made in phrenology. Doubtless there was greater profit from the lectures than from all the journals and publications combined (the *Constitution of Man* naturally excepted). Watson estimated that the sale of journals and the local memberships brought an annual income of only £400;[28] in individual cases, small fortunes were gathered on the lecture circuit. In Ireland Dr Carmichael, the leading phrenologist and President of the Medical Society, did not publish; but thanks to his lecturing neither did he perish. He readily gave £500 'to promote the cause of medical reform' — a generous sum by any standard, and almost entirely derived from phrenology lectures given over a two-year period. In America George Combe reckoned he earned almost $600 in five years from the sale of a few 'short tracts and pamphlets', but he made almost as much in a three-week tour of New England in 1839.[29] In England Simpson reported in 1836 that he was making fair profits from small but loyal audiences, and it is significant that George Combe married in 1833 and retired from law in 1836, fully expecting that his immense savings from publications could remain intact while ordinary expenses would be met by the proceeds of two or three annual lecture tours. Of course Combe gained much more from the *Constitution of Man* than he did from any other single source; and yet it was this volume, first appearing in 1828, which familiarised the public with the doctrines of phrenology and thereby set the stage for the lectures of the following decade.

In view of the success in the lecture halls and mechanics' institutes of Great Britain, it is interesting to contrast the approach of phrenology's most celebrated spokesman with that of a lesser-known but equally dedicated believer. George Combe never had long to wait for a speaking invitation; local phrenologists were constantly urging his presence as a means of general inspiration. In Richard Cobden's opinion, the very prospect of a Combe visit 'created excellent effects' and 'opened a new era for the mind in Manchester'.[30] By comparison, the speaking tours of Charles Bray of Coventry had less startling results. Converted to phrenology by George Combe, Bray (1811–84) spent the thirties and early forties testing the theories of Robert Owen and he took part in the 'Opening of the Millenium' at Queenwood in Hampshire. Bray's phrenology, like Combe's, was decidedly moralistic and didactic, and it rested on the deep foundation of human determinism which Bray elaborated as the *Philosophy of Necessity*. This, of course, was the pessimistic side of phrenology which did not generally appeal to audiences, but Bray saw no point in changing his message. Even in later years he stood by the austere philosophy which permeated such works as the *Manual of Anthropology* and *Force and its Mental Correlates*. Simpson regarded Bray's phrenology as the 'Calvinist branch of the science' for its severity of emphasis, but no one attempted to correct Bray, who was among the most influential phrenologists in the Midlands.

Unlike Bray, Combe insisted on certain conditions before undertaking a tour. The first prerequisite was a guarantee of a large audience. 'Never in any circumstances lecture on the mere personal assurance of your friends that you will have a large audience', Combe advised one associate. 'Men are sanguine, and premise on the buoyancy of their own hope.'[31] Unless a phrenologist was addressing a small group of artisans in their own institute, and unless he wanted to be mortified and disappointed, it was wise for him to determine on his own the number of persons likely to attend. Large audiences were important to Combe: they were the best way of 'knocking ridicule and contempt on the head'. Bray did not agree, feeling that phrenology was as effectively preached in small groups as in large lecture halls. Before going to a lecture, Combe also attempted to find out about ventilation and the rent of the hall, and as he often travelled with his wife and a servant he insisted on good accommodations and the use of a carriage.[32] Bray never demanded such things and toured frequently at his own expense, lecturing wherever there was the slightest interest. In the propaganda of phrenology, Bray's tireless reliance on lecturing rather than on tracts and pamphlets was born of the belief that people in the Midlands were not given to reading and must be converted by the spoken word.[33]

Charles Bray's preference for the lectern was common to phrenologists of the 1830s. They welcomed public speaking as an opportunity to change public opinion and to demonstrate their own personal importance within a world-wide movement. 'You will perceive that I am in the field again', wrote Dr Caldwell from America, 'endeavouring to enlighten ignorance and remove error, beat down sophistry and finish impertinence.'[34] While not every phrenologist set out to do as much as Dr Caldwell, there was no denying that the public were in a mood for lectures; for them it was a means of self-improvement and for phrenologists (like Dr Caldwell) it was an outlet for *self-esteem*. Public speaking was long regarded an Anglo-Saxon freedom, and now public lectures became an Anglo-Saxon duty. The lectures were not pitched in any particular direction. All might benefit in some way by the remarks. To the lectures came young and old, attracted, certainly, by the claims of the science and also drawn by a social compulsion to determine for themselves what was useful knowledge. Even youngsters went. Herbert Spencer first heard of phrenology as a boy in Derby. In his *Autobiography*, Spencer confessed that he was a bit repulsed at first by the sight of all the skulls which lined the front of the the speaker's platform, but this was probably a natural reaction for a lad of twelve. Once he had overcome his uneasiness Spencer became a 'believer' and he devoted much attention to phrenology throughout his adult life, although in later years he said he was merely a 'general adherent'.[35]

As phrenologists extended their lecture tours in the 1830s, Sir George Mackenzie wondered about the possibility of establishing a 'grand national society' for the propagation of the science, 'perhaps based along reformist-agitation lines'.[36] Mackenzie was ready to determine the reaction of leading Whigs, but on reflection he thought the project too risky. In the following year (1835), however, the society at Dublin urged the formation of a 'General Association of Phrenologists'. While Combe did not think much of the idea, a number of the local societies did. The phrenologists who (in their capacity as scientists) attended the annual meeting of the British Association in Newcastle in 1838 thought they would like to establish a 'separate section' to meet every year with the Association. Their proposal was rebuffed. But there was nothing to stop them from holding lectures and seminars immediately *after* those of the Association, which is precisely what they did at the convention in Birmingham in 1839. And when the Association met in Glasgow in 1840, the phrenologists again held their own 'section' and elected George Combe as its president. Apart from this show of independence as a science, the phrenologists accomplished very little. There were lectures, demonstrations of dissection, and reports on the progress of phrenology in Britain and Europe, but there were no agreed institutes or definitions of faith. The Phrenological Association warned the public against the unscrupulous quacks who perverted the science, but it was powerless to discipline or correct heretics within the ranks. The Association could do no more than identify and approve of those men who regarded phrenology as a scientific philosophy in the manner of Spurzheim and Combe.

The popularity of the lecture medium could not last. Its downfall as an effective instrument of propaganda was not the result of having too few to lecture around the country, but, in effect, of having too many. The lecture circuits were such a profitable business in the thirties that pretenders, unlicensed by the Association, also made a good living from the theories of Dr Gall. Those who respected phrenology as a full-blown mental philosophy were obliged to share their shillings with admitted quacks. Even in his bucolic post in Kentucky, Dr Caldwell complained that while phrenology's progress was still 'flattering', he was convinced the science would suffer 'temporary injury' due to the 'vagabond pretenders who are over-running the country, reading heads for hay'.[37] The injury, as Caldwell called it, was permanent and far-reaching. The 'vagabonds' confused the public; they helped to undermine phrenology's claim to be a serious branch of scientific inquiry; and they discouraged the 'orthodox phrenologists' from remaining in the field. As their audiences diminished, the lecturers increasingly fell back upon their original vocations for support. Watson returned to botanical research and spent his spare time dabbling in politics; Simpson withdrew to his educational experiments; others dedicated themselves to journalism, medicine or law. Even George Combe, while as convinced as ever and still lecturing, began writing on non-phrenological topics. The visual aids of phrenology lectures shared the same fate. The demand for graphs, skulls and plaster casts, so essential to the local museums and lecture-rooms of the twenties and thirties, were discarded or destroyed; surprisingly few of these items survived into the next century.

Still, there was no need for the prominent phrenologists of the twenties and thirties to go hungry through the forties. Theirs was a comfortable existence, often assured by incomes from other activity. George Combe certainly had more resources to husband than any of his colleagues. The *Constitution of Man* was still being sent to the press, and Combe regularly conferred with M. B. Sampson over the course of their investments in Britain and elsewhere. In America Combe easily survived the loss of a small fortune in railroad stocks, as did many British investors of the 1830s; the vocations and social rank of other phrenologists were guarantee enough of their survival. Just as many of them remained amateur scientists and amateur philosophers, so too did they remain respectable men, with the same connections to the respectable world of trade and professional services. For such men financial windfalls in the name of science were always possible. In 1840, for example, the Phrenological Society of Edinburgh almost obtained a bequest of £15,000 from Dr Roberton, a Scot who had died in Paris. Unfortunately, the sole executor of Dr Roberton's will was encouraged by the French government to block the transfer of the sum which, if delivered to the gentlemen in Edinburgh, might have subsidised a renaissance of their scientific efforts in Scotland.

In the quarrels and factions which existed at home, George Combe did his best to preserve the unity of phrenology as a moral philosophy. He was known as a peacemaker, able to minimise disputes and soothe the disputants. His letters were welcomed by phrenologists of all temperaments, including the pessimistic Charles Bray, whom they 'encouraged more than anything else'.[38] Combe's letters were also useful as propaganda, published in series as they were in the *Phrenological Journal*, or as individual tracts. By the mid-thirties, Combe occupied a position in Britain which no other phrenologist could match. He met frequently with professors, scientists and literary figures, rubbing shoulders with them at the Royal Society and receiving them as guests in his home. Even the old antagonisms between Combe and Francis Jeffrey, editor of the Edinburgh *Review*, were replaced by a new relationship of coexistence. Both men had to endure one another in their encounters in Edinburgh society, and so far did they leave behind their early battle stations that Combe was able, in correspondence and in polite conversation, to advise Jeffrey on the qualifications of various Edinburgh printers.[39]

At the same time, George Combe never exerted anything more than a spiritual influence over phrenology's scattered flock. It was not the sort of following which either produced or desired a commander, and Combe was notably reluctant to involve himself in the formation of a national organisation. Moreover, by the early forties, he was out of touch with events south of the border and he was forced to rely on two or three English correspondents (such as Sampson in London) for 'bits of phrenological news'. Combe's talents as an author, a peacemaker and as a socialite, while considerable, were insufficient to make up for the failure of the phrenological press, the phrenological societies or ultimately for the declining intrest in phrenology lectures; by himself he could never bring about the conversion of millions. To his friends he declared himself a realist because he admitted that it might take decades before the nation as a whole accepted the new philosophy. Phrenology, he said, was like a long distance runner whose only race was against time. Combe had clearly forgotten about all the hurdles along the way.

2

By the early 1840s, the downfall of phrenology as an intellectual study was apparent almost everywhere. As a method of transmission, the once popular lecture series joined the local societies and the journals as part of the *ennui* of phrenology. Even in the mechanics' institutes, where lectures on phrenology were so popular in the thirties, the enthusiasm had begun to fade. Those who continued to make the rounds, visiting local societies and mechanics' institutes, soon appreciated how uneven was the extent of conversion. The communities were small and scattered, their resources meager and reluctant. Phrenology was a badly proportioned state. Its most settled area rested among the Scottish societies, which outnumbered the English. By 1836 several communities had already disappeared from the national road map of orthodox phrenologists, and not many new ones were being established to take their place. In the mid-thirties Watson generously estimated the total number of paid memberships at about 900, and this number probably represented a high point in the history of the science.[40] Of course numbers alone cannot signify the real dimensions of phrenology's influence; many believed without having to pay annual dues. But, as Watson observed, phrenologists were more accustomed to meeting their opponents than their friends.

It was primarily in the cities and provincial centres that both friends and enemies were to be found. Phrenology thrived on the controversy which it generated among large numbers of professional people, clerics and academics. Debates involving public figures never failed to bring phrenology fresh attention and at least a few new devotees. In Edinburgh, to cite the most famous example, phrenology never did so well as it did in 1817, when Spurzheim spent several months in the city arguing with local opponents, and again in 1827–8, when Sir William Hamilton rather unwisely decided to confront the science. On both occasions the Phrenological Society of Edinburgh benefitted by the number of enquiries and by the applications for membership; the societies in small country towns never enjoyed this boost. In the industrial towns, too, phrenology managed to survive under the auspices of the mechanics' institutes. Several of them remained enthusiastic about phrenology into the 1860s; some continued to appoint 'curators' to watch over their collections of skulls and plaster casts.[41] Where these philosophical luxuries could not be afforded, the more usual connection was the presence in the mechanics' library of the *Constitution of Man*, a volume highly recommended to artisans by the Society for the Diffusion of Useful Knowledge.[42] In any case, whether among urban workers or among respectable tradesmen, internal complacency proved to be more harmful than outside criticism. 'We are all satisfied of the truth of phrenology'. wrote a member of the Dublin society, 'and unless something new came up, few of the members would assemble.'[43]

What did they expect to happen? For the smaller societies in the countryside, it was the hope that an eminent phrenologist would come and effectively destroy all local opposition to the science. If Simpson, Watson and the Combes had accepted half of the invitations which they received to make such crusades, they would have spent most of their time away from home. Phrenology relied too much on its transient spokesmen. It had been this way from the start, and the situation never improved. In Camrbidge, for example, generous arrangements were made for Spurzheim's visit in 1827 and the community (including the University) was fairly enthusiastic about the inevitable debates which his visit would generate. But the sensation came to an end when Spurzheim moved on. To remove the champion of a pseudo-science was to let the banners fall. It was not a problem of having sown seed in soil too rich; rather it was, as the *Phrenological Journal* bitterly suggested, a case of sowing the seed and then neglecting it.[44] Phrenologists were curiously indisposed to cultivating their own field. Too often they waited for the support of leading scientists or surgeons who were simply unwilling to take a risk on behalf of a credible but unproved mental science. This was true of Manchester, where the small circle of phrenologists decided that they needed only to obtain the endorsement of the city's senior physicians to insure the triumph of phrenology. Surveying a situation which he knew very well as a member of the local society, Cobden reported that there were many reasons why phrenology was not winning the minds of the Midlands. 'The primary one may be ... the fashionable timidity amongst the leading medical men and others, who although professing to support it privately, have not yet openly avowed themselves of this science.'[45] Cobden's comments were probably characteristic of the rank and file of most phrenological societies. They were men who respected the realm of science and who wanted, for various reasons, to become part of it; but could they be absolutely sure that they had entered that realm, or were only at the gates? Conscious of their own limitations, they waited for the reassurance of the experts.

Phrenology's dependence on ordinary men and its vain hope of support from professional quarters were both sources of weakness. Internally, too, the unproved notions of phrenology were bound to become a source of irritation. Ultimately the greatest source of peril to phrenology as a science was the divergent opinions among phrenologists themselves. While they were waiting for developments in medicine to confirm what Spurzheim and Combe had already told them, the phrenologists divided. The orthodox version of their creed allowed too much room for doubt, and it proved very difficult to keep even small societies on the straight and narrow path. Disputes flared within the most disciplined of the local societies. The society in Glasgow was certainly among the most orthodox; for many years it was ruled by A. J. Dorsay, an instructor in English at the city's High School. Dorsay was content with Glasgow, and he would suffer no division to arise as long as he presided over that community of phrenologists. But he presided too well. Dorsay was caesar; his realm was small and his senate conspiring. By 1840 the conspirators won the day, obliging Dorsay to withdraw. Sir George Mackenzie noted that the new regime in Glasgow was born of 'petty jealousies' and had turned from strict phrenological observance to outright heresy. 'They have started a new doctrine about the soul and mind as separate entities', wrote Mackenzie to Combe, and what was to be done?[46] In the end, nothing could be done. There was no one to call a conference, no way to heal the breach, and no course for anyone to follow but resignation. The cause of phrenology now splintered fast. Glasgow was a model for others. From the doctrinaire purists who kept the faith of Spurzheim and Combe (or who at least withheld their objections), power now fell to the reformists: to the experimenters, the mesmerists, and the clairvoyants, for whom phrenology was only a method of a much greater purpose. Even so, the decadence was long in the making. And, as many expected, it did not begin as heresy in Glasgow, but as a schism in London.

The London Phrenological Society began in 1823. Its first meetings were held at the home of its founder, Dr John Elliotson, who immediately sought the advice of George Combe and urged him to visit London. When Combe arrived in the spring of 1824, he found the condition of London phrenology 'more lamentable' than he had imagined.[47] The *Phrenological Journal* was not selling and, what was worse, the phrenologists were already quarelling among themselves for the leadership. There were several candidates and none of them was to Combe's liking. There was James Deville (d. 1853), the 'Great Apostle' of phrenology in the home counties; but he was a crude figure who spoke rough language and knew nothing of contemporary philosophy. There were two physicians, Donkin and Willis, men of some means, but they were not given to public speaking and, besides, they did not get along very well with their non-English comrades. Dr John Elliotson was the only bright light: Combe thought he was a man of good taste, education and ability, and noted that he kept a carriage. Unfortunately, Elliotson was not always even-tempered. By 1824 he had, through the force of his own irascible conduct, managed to split the London society into several factions. The so-called 'Zealots' supported Elliotson's claim to direct the proceedings of the society, directly or otherwise. Combe arrived just in time to enter the fray. Spurzheim learned of this in Paris; he wrote Combe to say that he thought intervention was probably a mistake. Indeed, if the London society was in such a mess, it would be better to concentrate on the building of a solid base for phrenology in the north. If that much could be done, Spurzheim was certain that the 'proper preacher' would later emerge in London.[48]

Nothing emerged from the capital but more trouble. Combe's intervention was resented, and he never again attempted to interfere so directly with the organisation of a local society outside the one in Edinburgh. Nor was the bad feeling caused by Combe alone. Spurzheim was not very good at following his own advice: he favoured the primacy of James Deville. In one respect, Deville, who was haughty and impossibly arrogant, was also indispensable to phrenology in London. He possessed in those early days the largest stock of visual aids: a vast collection of skulls, graphs and plaster-casts which Elliotson believed to be unequalled.[49] Deville was influential in other ways. As one of phrenology's first popular lecturers, he soon accumulated a fortune while his colleagues still had nothing. 'To whom do the public flock in

London?' asked Spurzheim rhetorically, already mindful of the answer. Clearly it was foolish to minimise Deville's merits; the man who 'has done more than all the other phrenologists together' had to be treated carefully.[50] Spurzheim's moderation was quite predictable, for Deville had already treated him carefully. He had taken plaster-casts of Spurzheim's head and entreated him to remain long enough to become phrenology's champion in London.[51] Elliotson would not hear of having so formidable a rival in permanent residence, and the old antagonism between Elliotson and Deville was now formally replaced by one between Elliotson and Spurzheim. After his Cambridge engagement, Spurzheim returned to London for a series of lectures. At the first lecture, Elliotson appeared and a dispute immediately arose over Spurzheim's honorarium and also over whether Elliotson should have to pay the admission charge (he probably expected a seat *gratis*, in recognition of his rank in London). Spurzheim returned the whole of the honorarium; Elliotson refused to accept it; and from that point on, the phrenologists who followed Elliotson in London were barely on speaking terms with their associates in the north.

In the case of the London phrenologists, therefore, schism preceded heresy. The estrangement of the mid- and late twenties essentially revolved around the problems of personalities and leadership — the sort of issues which Mackenzie termed 'petty jealousies'. Only in the next decade did noticeable variations in doctrine arise.[52] In fact, one important cause for the argument between these two main branches of phrenology was that their followers had little contact with one another. They relied on the post to convey local news and to explain problems in phrenology. As late as 1841 Mackenzie admitted that he had never met Elliotson.[53] And what the north did know about Elliotson made them all the more suspicious. He was much too delighted by his own reputation as an eccentric. He had discarded the silk stockings and knee breeches which still formed part of the standard dress of physicians and, in a time of clean-shaven chins, he was almost unique for his full beard. His wealth, his talent for publicity, and ultimately his good connections in London society all made Elliotson the only real hope of phrenology in the south, but there were cerebral signs in his large *combativeness* and *self-esteem* that he was capable of being 'harsh and unjust' with his fellows.[54] For the first few years — indeed, for the first decade — they detected nothing technically wrong with his phrenology. The confrontation with Spurzheim (and with Combe as well) changed all this. Elliotson had challenged two-thirds of the great trinity of phrenologists. Who remained but Father Gall? If Elliotson could only restore the science to the purity it enjoyed under its Founder, he would at a stroke advance the cause of mental knowledge and justify his own importance in the world of phrenology.

The restoration meant a change of emphasis. What Elliotson did was to give more attention to the surgical and anatomical aspects of phrenology than he gave to the philosophical and social applications which Spurzheim preferred. This did not imply that Dr Elliotson wished to minimise the 'practical character' of phrenology, for he believed that phrenology was eminently practical to the needs of medicine. His first objective was to finish the exploratory work of Gall and Spurzheim and to determine whether external conditions (e.g. physical pressure) might be used to activate the faculties. In time he discovered that his hunch was correct. Mesmerism was the answer. Admittedly mesmerism was not 'Gallism', for the revered founder of craniology had never foreseen all those curious techniques, some involving the use of ether, which Elliotson and his friends claimed as cures for various mental handicaps. But no matter. By his mixture of craniology with mesmerism, Elliotson at once revived the aura of progress in the pseudo-science, and he attracted many public figures to his side. Toulmin-Smith broke with Edinburgh phrenology in favour of London; so did Spencer. The two branches of phrenology were now further apart than ever.

Elliotson's move, which took him most of the 1830s to make, was practical and clever. He knew that it was not enough to continue his barrage against Spurzheim. He also realised that few Londoners would be attracted to the Old Covenant of Dr Gall. It was also clear to everyone in the mid-thirties that the Phrenological Society of London was, for all Elliotson's work, one of the weakest in the country. Richard Evanson, a secretary of the society in Edinburgh, wrote to Combe from London in 1837 that the society did not flourish in the capital, 'nor is it popular even with phrenologists'.[55] Watson sent a similar report, indicating that phrenologists had again divided into little cliques 'railling and ridiculing each other'.[56] In this situation Elliotson feared that his influence among scientists was beginning to suffer and he was also worried about his reputation as one of the capital's best physicians. He obviously had to do something to bolster his position, but making peace with Combe was out of the question. The decision to integrate phrenology with mesmerism — and with the by-products of mesmerism, such as thought-transference and mental manipulation — accomplished everything that Elliotson had intended. He salvaged the discoveries of Dr Gall in such a way as to show that they were truly useful after all and, more than ever before, he gained the fitful attention of London's medical and scientific community.

Apart from the personality of John Elliotson, there were other factors making for regionalism in phrenology. From an early date the reading material of the two sides was rather different. If he was to return even tentatively to the craniology of Gall, Elliotson could not very well favour the publications of Spurzheim and the Scottish phrenologists. Instead he promoted the sale and circulation of works by Londoners and he gave special preference to the tracts written by James Deville, which, curiously enough, he thought more useful than Combe's *Elements of Phrenology*.[57] This in itself was a wise expedient, for it placated Deville and at the same time encouraged him to supply the ingredients for another important regional difference. This concerned the accoutrements of phrenology. The plaster-casts which Deville produced in London found a market throughout the country, and he now made sure to label and locate the mental faculties according to Gall. While the variations were slight, the numbering pattern used by Gall and Spurzheim was not always the same. Here was an effective way of undermining Edinburgh's influence, for it created endless confusion among the members of those societies who knew not where to place their allegiance. 'The phrenology busts of Edinburgh and London are so differently numbered', complained Levison in Hull, 'that some information on the subject will much oblige us'.[58]
much oblige us'.[58]

It must be noted that the different texts by which London phrenologists learned their science, while helping to create a separate school of craniology, did not effect the sales of the *Constitution of Man*. Combe's great essay was popular everywhere in Britain. Elliotson's patronage and control tended to dominate only the output of lesser-known writers. Even so, the self-sufficiency of London seemed a backward step for a science as international and as intercultural as phrenology claimed to be. Believers outside the home counties put it down to the sheer truculence of Londoners. In Yorkshire and in Scotland particularly, phrenologists were almost as contemptuous of London as Cobbett himself. Could anything good come out of the capital? Elliotson's conduct was held to be symptomatic of the mental malaise; perhaps the original opponents of University College were right when they said that London was a place of wealth and not of learning. The *Phrenological Journal* saw it as one of the more unfortunate aspects of human nature: 'If Providence has constituted man solely as a labouring and hoarding creature, like ants and bees, then the citizens of

London exhibit nature in its most perfect state.'[59] Alas for the phrenologists! The older regionalism of custom, commerce and culture far surpassed any of the national traits they detected in the brains of men.

And yet, provided that Elliotson did not carry his phreno-mesmerism too far, it was not impossible to heal the breach. In 1841 Sir George Mackenzie proposed that an attempt be made to mollify the southern phrenologists. The British Association of Phrenologists might become truly national with the whole-hearted support of phrenologists from London and other societies which followed Elliotson. Mackenzie knew that many individual members of these local groups were already well-disposed to the Association, and his proposal aimed at the full and formal adherence of the societies themselves. His plan also involved an overhaul of the Association by provisions for the election of a strong executive body. For the first year, Mackenzie continued optimistically, it might be practical to elect a Londoner to head the executive – 'Elliotson himself, perhaps'.[60] Moreover, Edinburgh could well please the London phrenologists by taking a more active part in a campaign close to the southerners' heart: national education. Mackenzie's ideas were, in effect, proposals of peace and of reconciliation with Elliotson and his followers. But unfortunately they were not made from a position of strength. In the early forties mesmerism was fast making such inroads into phrenology that the science, with or without its Association, was in danger of being transformed. The impact was such as to confuse some and convert others. Dr Forbes, who held the chair of botany at King's College, London, told Combe that his 'faith in the testimony of phrenologists' was undermined after he had seen how easily some of their colleagues had accepted phreno-mesmerism.[61] Bray and Simpson cautiously approved of the mixture, and even Watson confessed that the alleged facts of mesmerism helped to fill a void in man's acquaintence with Nature and 'in our ability to explain how human beings can exert mental influence on each other by a mere word or glance . . .'[62] Only George and Andrew Combe continued to hold out against mesmerism, which they thought had no necessary connection with phrenology. As for expanding the British Association of Phrenologists, George Combe was quite indifferent. He avoided any personal involvement in the question by leaving for a long tour of Germany.

The advent of a rival journal in London was the last straw. It confirmed the phreno-mesmerists as schismatics and as heretics. In 1843 there appeared under Elliotson's direction *The Zoist: a Quarterly*

Journal of Cerebral Physiology and Mesmerism. At first Combe brushed aside the effect of this and similar journals, such as the *Phreno-Magnetic Indicator* published in Nottingham, but within little more than a year the danger to old-style phrenology was all too evident. The *Phrenological Journal* lost a quarter of its remaining subscribers — a defection it could ill afford — and when the *Journal* collapsed in 1847, *The Zoist* was still on hand to record the event in a bitter article.[63] At the centre of *The Zoist's* phreno-mesmerism was the charge that the psychology of Combe and Spurzheim had not moved with the times. Gall was treated with devotion as the true originator while Spurzheim was ignored. The general tone of articles in *The Zoist* was notably more medical and surgical, and the word phrenology was avoided. Elliotson had, of course, to live up to his name as a medical pioneer; he had been among the very first in Britain to use the stethoscope and now his phreno-mesmerism was proclaimed as a new and startling development in the field of mental physiology. To his defence came Dickens and Thackeray with their warm praise for his professionalism, and Herbert Spencer with his articles for *The Zoist*. By the publication of this quarterly, Elliotson showed that he meant to sever all remaining ties with the rest of the phrenological community, and it was but a short while after the appearance of *The Zoist* that the Phrenological Society of London was dissolved. It was an action which, once again, Combe regarded with complete indifference, believing that the London society had 'long ceased to experience any influence on public opinion, and did little to propagate the science among individuals'.[64]

The defection of former phrenologists to mesmerism was only one aspect of their drifting allegiances during the 1840s. Many, feeling that they understood the minds of men, committed themselves to various campaigns for social reform. As Ellis and Simpson did in education, these phrenologists brought with them their notions of human nature and responsibility. Others who entered the cause of social reform sought a newer psychological foundation to replace that of phrenology. Elliotson, Charles Bray and Harriet Martineau thought that phreno-mesmerism was the new covenant; Sampson and Dr Epps searched longer and eventually settled into the ranks of the English Homoeopathy Association.[65] In Dublin Archbishop Whately strayed equally far and became interested in clairvoyancy and 'somnabulic experiments' which he thought of great potential value to medical reform. All over Britain, many whom George Combe long counted as convinced phrenologists had each gone his own way. Simpson's 'fading enthusiasm' caused him to begin lecturing on non-phrenological subjects, and the cruellest blow came with the apostacy of Hewett Watson. The management of the *Phrenological Journal* was a trying business which soon convinced Watson that the science was not thriving. He first expressed his doubts in 1838, when he complained that he was unable to persuade good writers to contribute to the *Journal*. After less than three years as editor, Watson was anxious to return the job to the care of Robert Cox.[66] By 1840 Watson confessed himself a sceptic who could hardly believe anything of phrenology; he suspected that the shape of the brain was indicative of something, but he doubted phrenology's inferences. Increasingly he preferred to remain on the sidelines.

And so Combe was practically alone with delusions of orthodoxy. As far as he was concerned, the ambitious answers of phrenology still stood and the novelty of the science was undiminished. Why then should it decline? Combe attached most of the blame to the majority of ordinary phrenologists who knew too little of their science — or of one another, for that matter. He thought public ridicule had caused many of them 'to wither away'.[67] It was pointless for Watson to ask why no-one ever 'seceded from geology or from other sciences of equally recent origin' because Combe saw nothing intrinsically wrong with phrenology.[68] He was blind to the meaning of the misunderstandings which always plagued phrenologists, and he did not understand the real significance of Spurzheim's early warning against 'dogmatism'. The queries on various points of phrenology which were addressed to Combe in the 1820s were still being asked in the 1850s and this, too, had little significance for him. Embittered by the flight of his colleagues, Geroge Combe counted on the wisdom of future generations to vindicate his mental science. He consoled himself with the thought that no one remembered the names of those jealous and dishonest men who opposed Newton and Galileo; and so would it be, at some distant date, with the enemies of phrenology.

The decline through the 1840s went unchecked. There was nothing the old believers could do, and younger enthusiasts did not emerge to help them. The novelty of phrenology had worn away, and it was increasingly regarded as a form of entertainment. By the 1850s everything was left to the 'head reader'; only he (or she) could interpret the mystery of the bumps. The client was expected to know nothing about the brain, about its role in psychology, or about man's place in creation: the dominant style was now a variety of fortune telling. The more eccentric the head reader, the more mysterious the judgment. Continental phrenologists made the most of the situation: ironically they won the attention of common folk just as Spurzheim had failed to win a lasting victory over the learned. One such figure was Madam Cubi who, alive and well in London, described herself as the 'Spanish martyr of phrenology' because of the official opposition to the science in her own land. While she anxiously spoke her mind to William Ellis and Dr William Gregory, most of her associates from the continent avoided the older representatives of phrenology in Britain, with whom they had little in common. But whether foreign or homegrown, the head readers hastened phrenology's demise. Intelligent and respectable gentlemen found it almost impossible to accept a science now preached by scoundrels and money-grubbers. From his desk at *The Times* in London, Sampson politely explained the difficulty in rekindling the interest of intelligent people in phrenology. 'The fact is', he wrote to Combe after the latter's persistent requests, 'that there is a general rule to exclude anything that may be doubtful, and such words as "phrenology" and "homoeopathy" at once arouse this vigilance.'[69]

The cautious papers need not have kept watch. Phrenology's force was spent. The local societies had declined and the phrenological press was no more. Those who 'read heads for hay' replaced those for whom phrenology was both a living and a moral philosophy. Disenchanted believers either returned to the meetings of the literary and scientific societies or dabbled briefly in the newer fringe sciences. There was, however, yet another factor in the decline of phrenology: one of which nothing has been said so far. This was the persistent opposition of 'Old Religion'. It was easy for George Combe and his friends to say that the traditional belief in sacraments, miracles and revelation was an obstacle to knowing the true constitution of man, and they were certainly quick to denounce the 'narrow, sectarian demands' made by one group of Christians upon another. And yet the embroglio of religious debate proved hard to avoid, for there was in the mind of every man the fourteenth faculty, called *veneration*, which no good phrenologist could possibly ignore.

Notes

1. Spurzheim to Combe, 28 July 1823. NLS 7211/87.
2. Spurzheim to Combe, 13 March 1821. NLS 7207/76.
3. *Journal*, II, no.6 (1825), pp.241–3.
4. [John Epps] *Diary*, p.131.
5. *Journal*, introductory statement, I, no.1 (1823), p.vii.

6. Watson, *Statistics*, p.112f.
7. An account of the activity of one society is given in the *Journal*, I, no.3, pp.487–8. The article mentions visits made by German 'craniologists' and a phrenological discussion of Shakespeare's character Iago.
8. Watson, *Statistics*, p.136.
9. *ibid.*, p.115.
10. Watson to Combe, 3 January 1838. NLS 7248/137.
11. Combe to Watson, 1 March 1836. NLS 7386/508.
12. *Journal*, I, no.1 (1823), p.xiii.
13. Combe to Levison, 5 July 1829. NLS 7384/281.
14. Watson to Combe, December 1834. NLS 7233/183.
15. *Journal*, X, no.50 (1836), pp.301–37.
16. Combe to Levison, 5 July 1829. NLS 7384/281.
17. J. T. Smith to Combe, 12 January 1837. NLS 7244/12.
18. Watson, *Statistics*, p.15.
19. Watson to Combe, 18 November 1839. NLS 7252/152.
20. Watson to Combe, 20 December 1837. NLS 7244/82.
21. Watson to Cox, 23 September 1840. NLS 7257/141.
22. A handbill notice on Smith's lectures, dated 1838. NLS 7247/72.
23. Combe, *Lectures on Popular Education*, p.119.
24. R. S. Watson, *The History of the Literary and Philosophical Society of Newcastle-upon-Tyne* (London, 1897), p.232.
25. Combe, *Lectures on Popular Education*, preface, vi.
26. *Morning Chronicle*, 2 January 1815.
27. *ibid.*, 6 January 1815. Those who were not regular subscribers paid £1 11s 6d.
28. Watson, *Statistics*, p.221.
29. Nahum Capen to Combe, 17 March 1837. NLS 7242/48.
30. Cobden to Combe, 7 October 1836. B.M. 43,660/3 (C.P. XIV)
31. Combe to B. Hayden, 11 May 1837. NLS 7387/392.
32. Combe to J. Adam, 28 September 1836. NLS 7387/138.
33. Bray to Combe, 14 January 1848. NLS 7289/104.
34. Caldwell to Combe, 14 September 1835. NLS 7234/83.
35. H. Spencer, *Autobiography* (2 vols., London, 1904), I, p.228.
36. Mackenzie to Combe, 19 November 1834. NLS 7233/20.
37. Caldwell to Combe, 14 September 1835. NLS 7234/84.
38. Bray to Combe, 8 February 1853. NLS 7331/81.
39. Combe to Jeffrey, 24 April 1835. NLS 7386/287.
40. Watson, *Statistics*, p.220.
41. Mabel Tylecote, *Mechanics' Institutes of Lancashire and Yorkshire before 1851* (Manchester, 1957), pp.175f and 182.
42. *Manual for Mechanics' Institutions*, ed. B. F. Duppa (London, 1839), p.170.
43. Watson, *Statistics*, p.124.
44. *Journal*, X, no.53 (1837), p.706; also Combe to Carmichael, 17 March 1829. NLS 7384/217.
45. Cobden to Combe, 23 August 1836. B.M. 43,660/1 (C.P. XIV).
46. Mackenzie to Combe, 15 June 1840. NLS 7256/19.

47. George to Andrew Combe, 29 April 1824. NLS 7377/58.
48. Spurzheim to Combe, 13 May 1824. NLS 7214/59.
49. Elliotson to Combe, 23 July 1824. NLS 7213/16.
50. Spurzheim to Combe, 5 December 1826. NLS 7218/97.
51. Spurzheim to Combe, 13 August 1824. NLS 7214/62.
52. In *Phrenology, Fad and Science* (p.173f) Davies emphasises the 'disasterous schism' over naturalist philosophy which separated London phrenologists from the rest of the country in the 1840s. This was merely one aspect of the division which began almost twenty years before.
53. Mackenzie to Combe, 13 April 1841. NLS 7261/23.
54. Combe to Watson, 12 November 1837. NLS 7387/472.
55. Richard Evanson to Combe, 10 February 1837. NLS 7242/140.
56. Watson to Combe, 30 March 1837. NLS 7244/62.
57. Elliotson to Combe, 23 July 1824. NLS 7213/16–19.
58. Levison to Combe, 22 June 1827. NLS 7219/123.
59. *Journal*, VI, no.25 (1830), p.530.
60. Mackenzie to Combe, 13 April 1841. NLS 7261/23.
61. Forbes to Combe, 4 February 1843. NLS 7268/71.
62. Watson to Combe, 7 January 1843. NLS 7270/105.
63. Combe to Dr Hirschfeld, 16 December 1844. NLS 7390. *The Zoist* delighted in the 'premature decay' of its rival journal in an article in vol.V, no.20, p.421f.
64. Combe to Atkinson, 1 June 1846. NLS 7390/402.
65. Epps and Sampson were committee members of the Association and sent Combe a nicely printed advertisement of its meetings and activities. NLS 7277/48 (1845).
66. Watson to Combe, 9 August 1840. NLS 7257/134.
67. Combe to Simpson, 20 February 1843. NLS 7388/661.
68. Watson to Combe, 14 December 1853. NLS 7337/44.
69. Sampson to Combe, 28 September 1848. NLS 7297/39.

Chapter VI The Ungodly Error

When compared to their colleagues on the continent, British scientists were an unusual breed of men. Throughout the eighteenth century they refused to join the attack which scientists elsewhere had launched against traditional religion. Not that the British remained firm believers, untouched by the contagion of doubt; Hume, in the middle of the century, had seen to that. But it is remarkable that for the most part men of this profession in Britain should have entertained their doubts so quietly. Unlike the French they had little to gain by aggressive hostility to Christianity. Instead they continued to follow the example of Newton and attempted to prove whatever they could of traditional belief. Like Erasmus Darwin, they might question various items of Revelation, but they could not bring themselves to abandon theism. Like Joseph Priestley, they might annoy churchmen by their radical politics, but generally they preferred to maintain what one writer has called the 'holy alliance between science and religion'.[1] Most scientists assumed that their investigations were a part of 'natural philosophy' in which Providence had a definite role. Priestley was not the only scientist who thought himself more a theologian. It was only towards the middle of the nineteenth century that gentlemen realised the divergence of the two professions; until then the men of science and religion in Britain enjoyed a notably happy partnership.

Initially the British phrenologists fit into this pattern. As geologists, botanists, surgeons and physicians they believed that the discoveries of science were compatible with man's true religious needs. Their psychology compelled them to respect the Divine Plan as it was expressed in the laws of Nature. Indeed it was the custom of phrenologists to spell Nature with a capital 'n', just as many of their contemporaries did because it was in Nature (rather than in Holy Writ) that they found the most convincing proof of God's existence. Plentiful as the evidence was, the most startling single item of Creation was the brain itself. This sophisticated mechanism was implanted in man to lead him through Nature to Nature's God by the operation of the religious organs of *comparison, causality* and *veneration*. Together they produced a religious impulse, a 'feeling of devotion and gratitude' whenever man recognised his dependence upon the First Cause. To deny this facet of mental activity was to propose atheism, which was one of the criticisms phrenologists made of Robert Owen's *New View of Society*. According to the *Phrenological Journal*, the 'adoration of the Creator is not only the most rational but the most delightful exercise of the soul; and to found a system of society in which it has no place is to run counter to Nature and to despise the lessons contained in the history of the world.'[2]

The *Journal's* choice of words is revealing. There is the respect for Nature and for natural practices in society. There is the invocation of natural history in the broadest sense, the history of all creation, independent of cultural or doctrinal rubrics. And above all there is the belief that the 'adoration of the Creator' must always be guided by reason. Everything connected with religion must be rational. It was nonsense to talk about sin and depravity; if the religious faculties of the brain were inactive compared to others, 'sin' was only a malfunction of reason. The head of the greatest sinner was expected to show a severe deficiency of the reflective and moral organs and a predominance of the animal sentiments. Sin was not the result of weakness in the face of temptation but rather the inability to distinguish between right and wrong. The failure of religion was explained in the same terms of rational power. If the faculty of *veneration* was excessive and unchecked by the force of other organs, fanaticism of one sort or another was likely to result. There was no other way to explain why in Ireland the 'peasants worship the pope as the Egyptians worshipped crocodiles'; carried to excess, the irrationality of Old Religion was a cause of national madness.[3]

Their concept of rational religion, together with their respect for the natural laws, made most phrenologists reluctant to say very much about the First Cause Himself. Precise definitions could lead to sectarian controversy, which phrenologists wanted to avoid at all costs. The easy way out was an enlightened and generous latitude in religious matters. As long as people had the sense to admit that the First Cause must, almost by definition, be all-knowing and infinitely rational, what else mattered? This attitude brought leading phrenologists to the non-denominational theism which had been so fashionable among scientists of the last century. George Combe thought that the 'great body of educated men' must be like himself and his friends: deists of no particular stamp or perhaps Unitarians, who equivocated about dogmas 'out of fear of the fanatics'.[4] This good-natured ambiguity was common to phrenologists, but it was often a facade. Privately they entertained a wide assortment of religious doubts. Privately too, they resolved their problems in the strict monotheism of the Unitarians and were inclined to regard Christ as a man only, 'an admirable moral teacher' whose doctrines were harmonious with human nature.[5] It is no wonder that phrenologists, particularly in Scotland, were on friendly terms with Unitarians and had no scruples about donating to Unitarian building projects.

Religious broadmindedness came rather easily to phrenologists. They were born and bred in a variety of religious backgrounds and they did not all spring, as their enemies imagined, from the left-wing of the Christian spectrum. It was entirely possible to attend either church or chapel and to believe in craniology as well. The phrenologists knew that religious practices must differ from one country to another and even within a society because religious observance was a personal matter, determined largely by the cerebral state of men. Phrenologists were therefore scattered like their compatriots over a variety of churches. If they did express themselves publicly on doctrinal issues, there was no guarantee of unity. It depended on the phrenologist one might read; one anonymous work stressed the 'vital relationship' between man and a Creator who animated all things, while another questioned the certitude of afterlife.[6] And yet, while their affiliations were certainly diverse, in practice the majority of phrenologists were probably only nominal Christians. The only religious question of the day which roused their interest (and brought them to the speaker's podium) was the problem of toleration. Phrenologists thought that religious bigotry was the 'great moral disgrace of the age' and they advocated the widest possible toleration. While they were sincere in this (for good cerebral reasons), their position was also influenced by their political and social ideals. Their support of Catholic emancipation, for instance, involved no compromise of doctrine on their part, and they realised that their support of the Catholics was a way of striking the more dangerous Evangelicals, who were known to oppose phrenologists as directors and teachers in a number of schools.[7] In positive terms, therefore, what linked all phrenologists was their insistence on complete religious liberty; in negative terms, what united them was a hatred of sectarian feeling and legislation, which could be directed equally well against religious minorities or the advocates of a new branch of science.

Because of their own relative indifference phrenologists quickly tired of sectarian quarrels. They were not likely to appreciate 'enthusiasm' from any quarter. They were critical of 'narrow and illiberal policies' wherever they found them. The phrenologist-scientist Samuel Brown, author of the *Lectures on Atomic Theory*, declared that no group, from the Catholics to atheists, was less sectarian than any other. He characterised his own religion (in which there was plenty of room for phrenology) with the word 'independence', and he thought that every individual must be free to interpret Christianity and the Scriptures as he saw fit.[8] There could be no monopoly of religious truth and all sects were to blame for the bigotry and superstition which existed among Christians. Sir George Mackenzie blamed the Anglicans for not confronting the peril of popery when it arrived in the form of Oxford Tracts; he also blamed the 'Calvinist bigots' (a favourite phrase of Combe's) for having pushed people in the direction of romanism in the first place.[9] William Ellis thought the Evangelicals were responsible for the spread of popery; suspicious of their social and political activity he advised that they should be 'treated like children and restrained from mischief'.[10] John Epps decried the rituals of 'outward devotion' which effected most groups and which excited sentimental feelings while they depraved the intellect.[11] George Combe was so exasperated with the endless and vicious disputes among British Christians (particularly over the education issue) that he threw up his hands in despair and complained to Horace Mann that 'the Church of England is going back to Popery, and the Church of Scotland... to the Calvinism of Knox combined with the intolerance of the Popes'.[12]

For all this criticism, British phrenologists still looked upon the religious situation with a certain complacency. They suspected that toleration was a much more precious commodity on the continent than in their own land; and while they might favour Catholic emancipation at home, they were well acquainted with the record of princes and prelates who tried to suppress opinions (including those of phrenologists) in Europe. There were even times when phrenologists posed as defenders of English protestantism which, for all its foibles, was a marked improvement upon Romanism. On one occasion the *Phrenological Journal* corrected the 'irrelevant and puerile generalisations' of a phrenologist who equated the established church in England with Roman priestcraft elsewhere.[13] The equation was improper because the religious views of the British people reflected, by and large, the superior qualities of the British brain. This was the basis for the phrenologists' optimism and smugness — no matter how much they might complain of sectarian squabbles. British Christians had nothing comparable to the *Roman Index*, in which the works of continental phrenologists appeared. In fact the *Phrenological Journal* thought that the enmity of 'Roman priestcraft' was a sign that their science must be worthy of serious attention and sympathy everywhere, and it proudly published (in the original Latin) the papal bull proscribing recent French and Italian essays on phrenology.[14] That was proof enough, if any one really wanted it, of the modernity and true value of the new mental science.

The men of Old Religion were not convinced. For more than a century they had encountered these arguments and they remained suspicious. They had heard Creation described as a great and perfect Machine; they had known philosophers to praise human reason at the expense of Revelation; they wondered whether the demand for toleration was a cover for irreligion; and they had even known atheists to say that the religious climate of Britain was far preferable to anything in Europe. The fact that phrenologists fit so comfortably into the pattern of old-fashioned deism did not remove them from the scrutiny of old-fashioned Christians. Nor was suspicion reduced when Combe announced, in the opening pages of one edition of the *Constitution of Man*, that phrenology did not 'directly embrace the interests of eternity' and that the hereafter was uniquely the concern of theology, 'a different line of investigation'.[15] Far from calming clerical objections, the concession smacked of irreligion. Did it imply that these phrenologists, these so-called men of science, were not concerned about eternity and their just reward? Did it not suggest that theology was merely another course of study, on the same level as chemistry or geology, still unfolding and unfinished? It was not the first time such questions had arisen. The more closely some men examined phrenology, the more anxious they became: the challenge to Old Religion was unmistakeable, and as far as they were concerned no one – not even the 'Christian Phrenologists' – could disguise the danger which Gall and Spurzheim and their British friends had set in the way of true Christians.

1

In their own publications, the phrenologists often referred to the charges made against them. As early as the 1820s they began their essays, almost automatically, with a denial of atheism, materialism and fatalism (usually given in that order). The *Phrenological Journal* followed this course, and so did Sidney Smith in the first chapter of his *Principles of Phrenology*. But anticipating the attack rarely serves to prevent it. Moreover, it was quite difficult to defend the orthodoxy of phrenologists or to deny their irreligion when they themselves were involved in so many different projects and committed to so many different ideas. It was all very well for the *Phrenological Journal* to denounce what it regarded as the atheism of the Owenite movement; but churchmen in later years saw this move as something of a red herring when they were reminded that John Epps, Charles Bray and other phrenologists had once been, or still were, Owenites. Phrenologists denied atheism first because they knew that it was the most serious accusation which might be brought against them and they were aware of the social consequences of such a charge. They also realised how inaccurate it was. No leading phrenologist ever identified himself as an atheist, although several of them privately confessed to being agnostics. They found a position of complete scepticism intellectually more honest (if not more respectable) than outright atheism.

Phrenology had a contribution to make to agnosticism. In the past sceptics might have hoped to escape the traps of religious debate by saying as little as possible about God. Most leading phrenologists were equally evasive, and they found how useful certain expressions were, either for putting off an unwanted discussion or for hiding their own doubts and confusion. To refer to the 'Supreme Agent' or to the 'Author of our existence' or simply to 'Providence' created a suitably lofty and reverent tone but otherwise gave little away. Behind the gracious language, the Combes, James Simpson, William Ellis and Dr Carmichael lived daily with their doubts, just as their clerical opponents suspected they did; and in corresponding with each other over the years they realised the impossibility of what Old Religion taught about God and man. Their acute understanding of the human mind made it increasingly difficult for them to accept the supposed likeness of God and his creatures. How could men, with all their cerebral deficiencies, in any way reflect the wisdom and perfection of an All-knowing God? If such a Creator existed, obviously his mental qualities did not correspond with man's.[16] It was hard for phrenologists to believe, for example, that the organ of *destructiveness*, the physical embodiment of evil, should exist at all in the Divine mind.[17] This dissimilarity produced two dreadful alternatives. The first was that God has a mind so totally different that it is completely unknowable in its operation, its desires, and its plan for the human race; the second alternative supposed that the Divine mind *was* similar to the human, and therefore must include a number of imperfections and limitations — which in turn meant that God was neither all-knowing nor omnipotent.

The most fearless exponent of this agnosticism was Hewett Watson. When it came to denying the possibility of extra-mental ideas —

particularly religious ones — Watson was the most genuine sceptic and in 1843 he paid the consequences. He had hoped for an appointment to the chair of botany at King's College in London, but he withdrew his application 'on the understanding that theology and phrenology were both against me'. Watson realised that the college charter required professors to be Anglicans, but he could not meet this condition because he could neither deny or prove church doctrines. 'Thus I am neither a churchman nor a dissenter', he wrote to Combe; 'I am a nothing.'[18] How many other religious 'nothings' were there in the ranks of phrenology? If one were to judge from all those who confessed their disbelief and confusion to Combe, or those who reported to being victimised for their doubts (as was the musicologist Joseph Mainzer), it would seem that the sceptics were numerous indeed. To cast off traditional doctrines about the Almighty did not necessarily mean that phrenologists embraced atheism; many of them, like Watson and Ellis, were either confused or indifferent.

In any case, it was not easy to make the charge of atheism stick. The estrangement of phrenology and Old Religion was more readily proved on older and more familiar grounds. The most common objection was that phrenologists preferred a less supernatural religion and that they were attempting to give new life to the deistical and nondenominational notions of the past century. The objection was well-founded. There were few phrenologists who did not want to trim away the mysteries of Christianity; and they admitted it. Their goal was to show that man, through the possession of special mental qualities, was the apex of known Creation and that he could well have a 'higher destiny' than all other living creatures.[19] This made religion a more human phenomenon: more the product of human feelings and sentiments than of mysteries, miracles and half told stories.

If religion was to be less supernatural and more agreeable to all men, the first and most obvious casuality was Revelation. While phrenologists did not rule out the possibility of communication from on high, they detected many serious problems with Holy Writ. In practical terms it was too exclusive. It was unaccountably delivered to a few individuals who, acknowledging that these stories and lessons were only selective, reproduced them in manmade ink and parchment. The copying and recopying was productive of little more than error, and when preached by dogmatists like Calvin the Scriptures became (in Watson's opinion) a cruel joke on mankind. It was unfortunate that Christianity had to rely so heavily upon the gospels, for they created almost as much confusion as they did confidence in spiritual matters. The miracles were an awesome stumbling block. How was a good phrenologist, interested in human physiology, to explain the Laws of Health to an audience which took literally the account of Christ's raising the widow's son from the dead? And why had Jesus himself not explained the Laws of Health in order that no other widows would be deprived of their sons? Nor was this the only opportunity which Christ had missed. Combe and Cobden wondered whether Christ should not have encouraged temperance at the marriage feast of Cana by refusing to change the water into wine.[20] Phrenologically it was now possible to explain the origin of many of these stories. The heads of middle-eastern people revealed an excessive development of *ideality* and *imagination* and a relatively weak growth of *causality*; the result was that they were given to day dreaming and hallucinations; they believed what they wanted to believe. The great distance of time and the unreliability of the gospel accounts unfortunately did not permit scientific scrutiny of the miracles, but the mental characteristics of the people of Palestine were still visible. Alas! the narration of the gospels over hundreds and hundreds of years had wrought harmful effects on Christians; by perpetuating the tales of miracles the Scriptures had transmitted to the European mind a somewhat underdeveloped *causality* and a marked tendency to superstition.

Admittedly, the usefulness and veracity of Holy Writ was a point of debate among phrenologists themselves. Those who abandoned the Scriptures as a fundamental part of religion stubbornly refused to change their minds; in some quarters they were known as 'Combeists' because they shared George Combe's uncompromising view that the Bible bore the same relation to true religion as astrology did to astronomy. If the Scriptures were to serve any ethical purpose, they would have to undergo a radical and scientific revision, and Combe was certain that a 'new translation' would appear before the century wore out.[21] The guide to revision was phrenology itself: the Bible was to become a handbook of the temperaments and faculties — a manual of historic case studies in psychology. When this task was completed, the practical meaning of Scriptures would emerge, and the unwanted 'elasticity of the Bible text' (as George Eliot called it) would be replaced by the moral lessons of natural religion.[22]

The clerical enemies of phrenology therefore had no choice but to condemn the science for trying to substitute the laws of Nature for the holy commandments of the Old Testament. The indictment was very straightforward. The phrenologists were told that they, like so many others in recent times, thought that the stories of Holy Writ were too fantastic to be true, that they looked instead for a covenant with Nature which required no copying and no new editions. They imagined, wrongly, that the so-called 'laws of Nature' were perfectly and equally clear to all honest men. In short, the phrenologists, while perhaps well-meaning, were captivated by those very illusions which in the long run would abolish religion altogether. The *Constitution of Man* was only the latest source book of such illusions. One pamphleteer, in an 'Address to the People of Scotland', made plain his reasons for condemning Combe's work. What advantage could Christians possibly derive, he asked, from rejecting the Scriptures in favour of natural laws which no one had yet listed or classified?[23] To some anti-phrenologists this seemed as foolish as the care taken in ancient Athens to provide for the worship of an unknown god; to others it was more pernicious. They found bold statements of natural religion in the most unlikely places. Andrew Combe was a frequent offender: in his popular essay on child care he wrote that even the Almighty 'acts according to fixed general laws . . . ; and to disregard these His decrees, is as truly to rebel against His will as it would be to act in face of any of His written commandments.'[24] Similar statements were to be found in every edition of the *Constitution of Man*, and with every fresh edition Combe was censured for his attitude to the gospels. Never did he compromise his belief that a religion of Nature must replace that of revelation. It was his understanding that God required nothing in the way of prescribed worship but simply desired that man's *veneration* be exercised by respecting the visible 'plan of creation'. Indeed this was all God could properly do. He was not an autocrat but a constitutional monarch who observed human actions and did not govern the world 'in any practical sense'.[25]

The men of Old Religion were already well acquainted with these arguments. And from past experience they knew what the outcome of natural religion was likely to be. If one removed supernatural interference from the history of man, if one denied the miracles of Christ and the gift of Revelation, if one insisted (as Holbach had) that there was in fact nothing beyond Nature and that science alone could make man happy, then the whole basis of morality must also change. It was by now a commonplace – and a dangerous one, in the view of most clerics – that a person could be 'naturally moral'. The implication here was two-fold: a person might conduct himself in an entirely ethical manner, beyond the reproach of anyone in Christendom, while at the same time taking no part whatever in religious worship. Many phrenologists did just that. They believed that certain faculties of the brain, when unfettered by the sectarian doctrines and ceremonies of Old Religion, were able to distinguish between what was proper and what was wicked. Their objective was a code of conduct eminently sensible, rational and benevolent; morality must be uniform because goodness was universal and because the aspirations (if not the achievements) of the mind were universal. An instinctive, natural morality thus complimented a natural religion: the supernatural was severely trimmed away in both, and both were built upon what was knowable of the human condition. No matter how often clergymen had confronted these notions before, there was now a certain fullness to the natural morality phrenologists were preaching. It was as if the lessons of phrenology were, in themselves, a new convenant – didactic, moralistic and scientific all at the same time. Many were attracted to phrenology for that reason alone: Emerson declared that the chapters of the *Constitution of Man* were the best sermons he had read in a long time.

The anti-phrenologists were duly alarmed. It was exasperating for them to find people entangled in all the doubts which the previous century had entertained on the subjects of Revelation and supernatural religion; but for these phrenologists to lend their scientific arguments to the making of a new morality, to make a religion of their irreligion: that was devilish indeed! Already, in the highest professional circles, phrenology was having a religious impact of its own, and its progress was alarming. Of Dr Epps' essays on phrenology the *Monthly Gazette of Practical Medicine* commented that the writing was 'extremely beautiful' and praised the author for 'holding up phrenology as a moral light' and for showing its 'intimate connection with religion'.[26] There was little reason, in the *Gazette's* opinion, to consider the findings of Gall and Spurzheim subversive of religion, but clergymen -- Anglicans and Dissenters alike -- continued to worry and to publish their attacks. Before them was the spectre of a secular morality constructed by amateur scientists; as guardians of their flocks, they could never afford to let phrenology's claims go unanswered. The volume of printed material testifies to their fighting spirit. The tracts and pamphlets rejecting phrenology on purely scientific grounds probably reached a highpoint in the 1820s and early thirties and then diminished (both in number and emotion) as professional scientists lost interest in what Gall and Spurzheim had to say. Not so with the Christian gentlemen. Their verbal and printed onslaught began early, grew in intensity through the 1820s, and was sustained at fever-pitch through the thirties and forties. Theologians and obscure country pastors all felt obliged to warn their people of the danger. It was not so much phrenology's empiricism or its rationalism or its anticlericalism which upset them; they had detected these traits at the very beginning, as they polished the arguments which had brought them through similar battles in the past. No: the special danger of phrenology was that it combined all the old features of irreligion under the mantle of a new scientific philosophy which was understandable to all classes of men.

If a single word summed up the menace of that philosophy, the word was materialism. To measure the various segments of the brain was one thing; to say that they were responsible for different mental attitudes — or that they generated the human spirit — that was another. Materialism was a bugaboo which phrenologists faced in every audience and in the argument of every anti-phrenological pamphlet. As late as 1838, a writer in the Sheffield *Iris* was still trying to demonstrate the unchristian implications of phrenology: if the mind could not exist without the brain, then the mind was dependent on matter; therefore phrenology led to materialism. It was the same terrifying logic which, because it said nothing about the soul or the soul's relationship with the mind, had caused the Austrian government to forbid Gall's lecturing thirty years earlier. Spurzheim, by comparison, was never forced out of Britain, but he was obliged in public lectures to defend his ground. So often did he encounter the complaint of materialism that he decided to raise the question in his *View of the Philosophical Principles of Phrenology*. The answer was a rather unsuccessful sleight-of-hand. Spurzheim could not very well deny the materialistic outlook of his science, but he could try to make it seem less malicious than his opponents imagined. There were, he said, two types of materialism. One held that there was no God, that matter always existed, and that all phenomena were aspects of matter. The science of phrenology had 'nothing in common with this sort of materialism' because phrenologists admitted 'a Supreme Understanding, an all-wise Creator'. Moreover, Spurzheim continued, neither he nor Dr Gall had ever 'made any inquiry into the nature of the soul'.[27] Phrenology's materialism, in other words, was not the harmful variety.

For all its shortcomings, this was the only answer phrenologists could make. They might embellish the argument with a few interesting analogies, but essentially it remained the same. When the *Phrenological Journal* first appeared in 1823, it approached the problem of materialism by comparing phrenology with music. The musician depended upon his violin, a material object, to produce a series of notes, which were 'spiritual and abstract'. Similarly, ideas come from

The Ungodly Error

that material substance known as the brain. 'The brain and the violin are alike tools', the *Journal* concluded, 'and reference to them is essential.'[28] The analogy failed to move those anti-phrenologists who insisted (as one clergyman did) that the brain and the mind are actually independent of one another.[29] Phrenologists tried other comparisons. The *Journal* later explained that the mind was not the name given to any 'existent being'; it was only a convenient term, used to refer to various actions collectively. The process of thinking was like that of breathing: it necessarily had reference to corporeal functions and motions. The common denominator to all phrenological arguments on this issue was the evasion of anything to do with the soul. They all repreated Spurzheim's policy of disclaiming any authority in spiritual matters. They granted that the mind had no identity with the soul and that 'we must look to Revelation alone for any instruction about the nature and destination of the soul'.[30] That was perhaps a safer answer; it was clearly calculated to soothe religious criticism by separating religious considerations from the anatomical.

Unfortunately for the phrenologists, it had no such effect. Their apparent disinterest in the spiritual side of human existence was more suspicious than it was reassuring. How could these phrenologists claim to have a complete and true philosophy of man when they refused to come to terms with man's quest in this life for salvation in the next? George Combe thought this objection was 'futile and unphilosophical', and he warned anti-phrenologists from making 'imbecile and unfounded attacks' which would serve only to 'diminish the respect with which it [religion] ought always to be invested'.[31] Actually the attack was not unfounded at all. Phrenologists avoided making any judgments about the soul because most of them were not certain of its existence. In this respect they were (and were widely suspected of being) like Priestley, whose materialism led to a religion purified of 'corruptions'. The idea that the spirit or the soul of man was separate and immaterial from the body was considered, both by Priestley and the phrenologists, as 'part of the system of heathenism' which had contaminated Christianity over the centuries. But whereas Priestley had declared openly his disbelief in the soul (and criticised Locke for not having the courage to do so), the phrenologists were afraid to make that element of their irreligion known and denounced. If for the sake of argument they *did* admit the possibility of the soul's existence, they attached the proviso that 'both the soul and body are organs acted upon by an organizing law'.[32]

If the totality of man (including his soul, assuming he had one) was subject to natural laws, and if his feelings of devotion were produced by measurable parts of the brain, there was a question of human freedom. Was man a free agent in worshipping God? Here was another religious issue containing so many philosophical traps that phrenologists preferred to remain anonymous when they tackled it. One such writer, concerned with the 'moral influences' of his science, decided that man enjoyed the free exercise of his reasoning powers.[33] But what if, as the same writer asked, the faculties were somehow perverted? Did this not mean that an oversize *veneration* could compel man to a religious rigour beyond the prescriptions of any church? It was already the view of orthodox phrenologists, taking their cue from the *Constitution of Man*, that such an abuse was not only possible but a matter of historical fact. Phrenologists were thus left with an awesome variety of determinism; on both sides of the Atlantic they spent much time and energy wrestling with it. In America the phrenologist Isaac Ray frightened the editors of the *Christian Examiner* by an article in which he described the mind as 'a determined and original faculty'. While Ray's understanding of phrenology was far from correct (he should not have spoken of the brain as 'a' faculty), the mechanistic features of his craniology were not lost upon his colleagues. Several of them were Unitarians, and they regarded any sort of determinism as Calvinistic.[34] Isaac Ray's error was explained to him. He then retreated a little and emphasised that the religious faculties, like any other group of faculties, only 'disposed' or 'inclined' men to certain acts, and there was no need to worry if the faculties were part of a well-balanced mind. There was a desperation in this defence (which most phrenologists used at one time or another) because the last thing Ray wanted was to be accused of the same unyielding determinism which he supposed peculiar to Calvinism.

The same problem confronted the Phrenological Society of Edinburgh, where it eventually created a serious dispute over the question of man's free will. What Spurzheim had to say on the subject – at least in his published works – was not incompatible with orthodox Christianity; he endorsed Augustine's view that God does not impose necessity upon man and that human freedom presupposes the ability to will for or against something. To this formula Spurzheim added a phrenological postcript. The process of selecting a course of action involved the power to compare the alternatives: hence human will began with the reflective faculties, notably that of *comparison*. Spurzheim's conclusion turned out to be surprisingly agreeable to Old Religion: he held that cerebral determinism as such did not exist because man chose on the basis of personal knowledge what he thought good or desirable. Sometimes the knowledge was faulty; other times it was incomplete; at any rate, the use of the will was the business of individuals who acted for a plurality of motives and with different appreciations of the possibilities before them.[35] But in the privacy of their meetings, not all Edinburgh phrenologists accepted this position. A number of them, led by George Combe, had long nursed genuine doubts about the Christian teaching on the will, and by 1826 they were ready to accept the logic of phrenology on the subject.

As a basis for the discussions in the society, Combe arranged for the private and limited printing of an essay which he circulated among the members. He expected that some of the ideas which he set down in his essay 'On Human Responsibility' might disturb some of his associates, but he was hardly prepared for what one writer has called the 'general chorus of opposition' which actually occurred.[36] The essay was much too critical of Old Religion and of the doctrines which many of the Edinburgh members, as professional men, preferred to take for granted. Led by William Scott, one of the founders of the society, they objected to Combe's essay because it focused too clearly on the distinctions between rational and supernatural religion. Scott believed that Combe attached too much importance to the 'religious faculties' – so much, in fact, that man was not wholly responsible (and therefore not liable to punishment) for his misdeeds in this life. For Combe the Will was subsumed in behaviourist psychology; for Scott it remained a separate entity, in no way material, decidedly mysterious, and a gift of God. Scott readily conceded that it was no easy business to define Will; but of one thing he and his friends were certain: to deny the gift of the Will, as Combe appeared to do, was to question the generosity – if not the existence – of the Giver himself.

It is difficult to say whether Combe intended it or not, but in effect the 'Essay on Human Responsibility' was a trial balloon. Not a few of the ideas which divided members of the Edinburgh Society were to reappear two years later in the *Constitution of Man*. On both occasions the results were tumultuous, but for Combe there was no turning back. His doubts had finally brought him to a complete rejection of Old Religion. Several of his fellow members made or were soon to make the same journey; others felt that Combe was taking phrenology much further against religion than Spurzheim had intended. This was, in a sense, the most logical path for phrenology to take – if only some one pushed it in that direction. Until Combe came forward with his essay in 1826 the Society had studiously avoided committing themselves. They granted that phrenology was a materialistic philosophy, but they found ways around its being too deterministic. They earnestly hoped for more research to support Spurzheim's notion of a 'plurality of motives'. They had, in other words, shied away from the most frightening and the most logical inference of phrenology. Now the moment of truth had come. Most members remained in the Society; while they did not all agree with Combe, they saw no reason to leave. William Scott did not remain. He had desperately wanted to reconcile his phrenology with his religion (he was able to do so later, attacking Combe at the same time)[37], but it was now clear to him that Combe was against this combination. Scott's position is best illustrated by the advice which he gave to Combe at the end of the Society's discussions in 1826. 'You must not hold out what you have so often repeated to me,' Scott warned, 'that Christianity must fall unless it harmonises with your conclusions . . . such language is highly offensive to those who believe [in Christianity], and is calculated to excite the propensities [of others] against you.'[38]

With those remarks Scott took his leave both of George Combe and of the Edinburgh Phrenological Society. And he was not the only one. A few years later, another prominent member, the Reverend David Welsh, decided that on the basis of Combe's essay it was advisable to quit phrenology altogether. Welsh, who first suggested the founding of a society in Edinburgh, placed a much higher value upon his work in the Church of Scotland where (according to Hewett Watson) his orthodoxy was rewarded by a chair of religion.[39] And yet he had always appeared to be a genuine convert to phrenology. During the 1820s he addressed scores of letters to Combe, describing as accurately as possible the mental faculties (especially numbers 3, 6 and 14) of everyone he encountered. Before phrenology became such a preoccupation for him, Welsh had read widely in the sciences (although, as he confessed, without much guidance); and he thought phrenology by far the most interesting. For years Welsh considered Combe his best friend and it was through his influence that the *Literary and Statistical Magazine of Scotland* agreed in 1816 to publish Combe's first essay on phrenology. But by the early thirties his break with the author of the *Constitution of Man* was complete, both philosophically and socially. The materialism which Welsh had always recognised as part of phrenology had grown into a monster, hostile to Christian doctrines. Anti-phrenologists were naturally pleased by his move; it was, they said, the decision of an honest man who had discovered for himself the true meaning of phrenology. But if they expected many others to take the same course as the Reverend David Welsh, they were to be disappointed.

2

Mention has already been made of the 'Christian Phrenologists'. Unlike Welsh they believed that the teachings of Spurzheim were compatible with those of Christ. They thought that phrenology, like other branches of science, might endure religious opposition at first, but the opposition would not last. After a few initial murmurs Christianity had accepted the Copernican Revolution; to the Christian Phrenologists it was cheering to realise how over the centuries scientific research and Christian doctrines had adjusted to each other. True, the *Constitution of Man* emphasised natural religion and neglected the 'interests of eternity'; but that was not necessarily a declaration of war. According to one Christian Phrenologist, the new science meant no harm. 'To suppose that life and mind are the results of organism may be possibly absurd', he reasoned, 'but it is by no means impious, because it interferes with no religious dogma.'[40] Even so, the long and loud controversy over the *Constitution of Man* frightened at least a few phrenologists into what seemed a philosophically safer place.

The Christian Phrenologists were united by their enthusiasm for science and they were known equally well for their lectures on physiology and phrenology as for their Sunday sermons. The Reverend George Barlas was typical. He regularly travelled to neighbouring towns where he made phrenology the subject of public meetings.[41] In addition to itinerant preaching, there was an almost evangelical flavour to Christian Phrenology. There was less concern for the system of sacraments, the functions of priesthood and issues of church organisation; instead the emphasis was on vitality and improvement. David Clyne, an Edinburgh phrenologist, proposed a 'society for the practice of Christianity', assuming that phrenologists understood the mentality and spirit of religion.[42] Other phrenologists in the 1830s and forties were also attracted to the idea; among them was George Goyder, a Swedenborgian minister in Glasgow. Unfortunately his congregation did not share his enthusiasm for a more scientific Christianity, and (as he complained to Combe) they dispensed with his services because they thought him too devoted to phrenology.[43]

Whatever their backgrounds and plans, the Christian Phrenologists were inclined to believe in a religion reduced to nonsectarian principles. It is possible to say of the Christian Phrenologists what one writer has said of the Romantic Age as a whole: there was the disposition to think that all religion required was 'the application of common sense'.[44] They had no particular desire to add to the complexity of doctrinal religion, and they were more interested in adding to the social and scientific plausibility of Christianity. Their contribution was often unusual. One Christian Phrenologist, for example, demonstrated the devotion of early martyrs by an analysis of skulls brought from the catacombs. Were the religious and moral sentiments of the ancient Christian skull really more pronounced than those of pagan skulls? The Christian Phrenologist thought there *was* a difference, however slight; and no anti-phrenologist hurried to deny the superior qualities of the Christian mind. Besides, the objective of the Christian Phrenologist was commendable enough: he wanted only to preserve the old harmony between science and religion and to build a new Christian certitude with the peculiar methods of his own science.

But how exactly was phrenology to help out? Given the non-sectarian outlook of most phrenologists and the fundamentalism of others, it was fitting that they should have focused their attention on the Scriptures. To Henry Clarke, the author of *Christian Phrenology* and a minister in Dundee, the connection between mental physiology and the teaching of the New Testament was perfectly clear. From the Bible, that useful sourcebook of diverse conclusions, he catalogued numerous examples of *wonder, hope* and *love of approbation*, and his example was followed by Christian Phrenologists everywhere. The Bible thus became the most important of all protophrenological commentaries: it contained valid descriptions of human nature as well as rules for human conduct. The only thing the Bible did not do was to name the mental faculties which it described; this is where modern science came to the rescue. The woman who begged Solomon not to cut her child in half was said to be guided by a large *philoprogentiveness*; the Pharisees were cited for their inordinate *love of approbation*; Job was obviously directed by his *causality* and *hope*, Lazarus by his *acquisitiveness*, and Herod by his small *benevolence*. Of course no measurements of Biblical characters were actually available to phrenologists; they could refer only to stories which were, after all, divinely inspired.[45] So rich and complete was the Bible in its tales of mental torment and motivation that even the *Phrenological Journal* (still in Edinburgh under Combe's influence) favourably reviewed Clarke's work.[46]

The merger of phrenology and Christianity worked both ways. On the one hand phrenologists could say that their modern knowledge of the brain verified the stories of mental anguish, mental resignation, mental plotting and mental innocence all found in the pages of Holy Writ. On the other hand, the same stories were told in the hope that people might accept a measure of Biblical support for phrenology. It seems that Christian Phrenologists were quite sincere in their use of the Bible, but it was also a good tactic on their part. Even the Reverend David Welsh thought for a time that people would be more receptive to the science if phrenologists themselves occasionally drew their examples from the Bible. As it was, in the late 1820s phrenology bore the imprint which Spurzheim and Combe had given it: it was overwhelmingly concerned with the material world of anthropology and medical science. Welsh's apprehension was that phrenology should not be allowed to drift any further from the spiritual and supernatural concerns of man. He was well aware of the increasing number of letters which Combe received — often short anonymous notes — asking what was to become of Christian practices (such as prayer) if phrenology were true.[47] What Welsh and others hoped to do, therefore, was to restore a balance. The Christian Phrenologists wanted to show by their use of the Scriptures that they were acquainted with Christianity and that they accepted it.

The balance proved, in the long run, to be an illusion. Some Christian Phrenologists (like William Scott, formerly of the Edinburgh Society) tried to defend the general concepts of cerebral faculties and the localisation of ideas, while at the same time they disassociated themselves with the 'philosophical errors' of the *Constitution of Man*. Others, like Dr Epps, welcomed Combe's work and saw nothing irreligious about it. Still others decided that the *Constitution of Man* was entirely acceptable if interpreted properly: in America the Reverend Joseph Warne prepared a special edition for the use of New York public schools; the edition included a chapter of his own, 'On the Harmony between Scriptures and Phrenology'. Warne's argument was perhaps a bit too clever. He equated phrenology and Christianity for the guidance they attempted to give mankind. 'Phrenology', he said, 'teaches us the possibility of *individual improvement*, which the Scriptures term personal conversion.'[48] It was an interesting argument but it did not resolve anything. To speak of 'individual improvement' as the phrenologists did was to speak of human effort and human application whereas traditionalists preferred to think in terms of a sudden, free and quite mysterious intervention from Heaven. Nor did Christian Phrenologists gain much from their forbearance with fellow Christians. They never completely lost patience with the sects of which they were at least nominal members, but most of them were known to be as weary as Combe and Watson were of sectarian squabbles and rivalries. And if the Christian Phrenologist subscribed to only one-half the ideas of the *Constitution of Man*, he was still in trouble with clergymen who thought that man's 'first object is to glorify God' and not to 'study the elements of external nature'.[49]

While the stigma of materialism was not completely erased, Christian Phrenology was enough of a development to satisfy some and entice others. David Welsh was probably the only important member of the clergy to turn his back on phrenology; some of his colleagues continued to praise the Lord and preach phrenology in their spare time. The study of psychology was a matter of some import to the clergy as a profession and particularly to those aware of the contribution made by clergymen in the past. The ideas of phrenology, illustrated by Scripture and confirming God's goodness, attracted Christian gentlemen all through the 1830s and forties. Simpson reported in 1839 that the bishops of Durham and Norwich were attending a series of phrenology lectures in London. Whether they did so for amusement, when the Lords were not sitting, Simpson did not say; but they were described as amiable and genuinely interested.[50] It is unlikely that Christian Phrenology caused any great increase in the total number of phrenologists in the country, but it is significant that a few of the new societies formed in the thirties (notably at Bath, Aberdeen and Warrington) won the support of local ministers. The *Constitution of Man* did not drive them all away. And of all the clergymen who still wandered into phrenology, whether through scientific curiosity or the study of human morality, surely none was more remarkable than the Archbishop of Dublin.

Shortly before he was elevated to the archbishopric in 1831, Richard Whately (1787–1863) tried to determine the merits of phrenology. To George Combe he sent a plaster-cast of his head and asked for a frank analysis. To be safe, Whately made similar requests of other phrenologists in England and Ireland. All the evaluations seemed to point to exceptionally strong moral and reflective qualities (only natural in an Oxford professor and a rising clergyman) and Whately was satisfied. To those who knew anything about him, there was no denying Whately's *self-esteem*; he once wrote to Combe that if Robert Chambers should ever wish to investigate the Irish Problem he should come at once to consult the Archbishop. 'I know a good deal of the questions and answers', Whately assured Combe, 'and Mr Chambers may rely on my impartiality.'[51] Whately first became interested in phrenology as an explanation of mental attitudes, and he thought that even if it were someday proved that the mind had no relation whatever to the brain (a discovery which the Archbishop thought very unlikely), still, the phrenologists had devised the 'best nomenclature' of mental activity.[52] In fact, the more he read about the science, the more it seemed to coincide with his own approach to Christianity.[53]

Few archbishops have ever approached their duties or their religion as Whately did. At Oxford in the 1820s he declared his opposition to evangelicals and to erastians. He denounced Romanism and yet tried to remain on good terms with Newman and Pusey, both his students at Oriel. Harriet Martineau wrote that Whately's appointment (which came in 1831 as a reward for his political liberalism) caused almost as much surprise in England as in Dublin – and well it should have. Doubtless the public would have been even more astonished had it known of Whately's impatience with the treatment of Catholics in Ireland: his 'fondest wish', which he once intimated to George Combe, was that he would like to see all the cathedrals in Ireland burned down – those 'irritating monuments to injustice' which Catholics had built but could no longer use. Indeed, Whately considered the established church (in England as well as in Ireland) an open question, an issue similar to the Navigation Act or to the Corn Laws, deserving 'open and polite argument'.[54]

Phrenology came as a great comfort to Archbishop Whately. Long interested in the sciences and anxious to meet the needs of practical education, he found in phrenology not only the 'best nomenclature' of mental attitudes, but the most durable explanation for human morality. He drew a parallel, as most Christian Phrenologists did, between primitive minds and primitive religious practices, and the result was a vindication of Christianity. So persuasive was the argument that Whately developed it as a fitting example in the *Elements of Rhetoric*. His opening statement was the most reassuring part of the case: 'Nations of atheists, if there are any such, are confessedly among the rudest and most ignorant of savages.' Any people who respected their God or gods 'as malevolent, capricious, or subject to human passions and vices' were usually brutal and uncivilised. This being the case, even Moslem societies were thought better than their pagan neighbours, because at least the Moslems 'maintain the unity and moral excellence of the Deity'. And for the same reason Whately concluded that the nations of Christendom were undeniably 'the most civilised part of the world' because of their 'most cultivated and improved intellectual powers' and because they esteemed goodness (*benevolence*) more highly than did other peoples. The basis of Whately's argument is that premise to which phrenology added so much weight: man is a rational creature, but among men (in Whately's words), some are 'most rational and cultivated'.[55]

Whatley's use of phrenology as an anthropological buttress of

Christianity was by no means unique. Whether they followed Elliotson or Scott or Combe, phrenologists argued that Christianity (for all its defects) was still the reflection of a superior mentality. And having said that much they were almost obliged to remain Christians. But unfortunately Whately disappointed them. It was not that he played down his phrenology; he resorted to his 'preferred nomenclature' even when writing to non-phrenologists. The real problem was his reputation. Everywhere there was talk of his heterodox opinions. His firm belief in religious toleration was taken as a sign of personal indifference, his loyalty to Blanco White as proof of his Unitarianism. Christian Phrenology could hardly prosper when the Christianity of its best-placed advocate was a matter of general doubt. It was a most ironic situation. The phrenologists, long accused of godless materialism, now shared with Anglicans a certain embarrassment over Whately. 'At least the Archbishop is not an infidel', Combe remarked in 1846; but that was little consolation. Phrenologists came to realise the vanity of their hopes in having the patronage of an archbishop.

Or were they expecting too much of Whately? Had he been more straightfoward a personality — had he been more careful in the public eye — the result would probably have been the same. The Christian variety of phrenology could not save the science from its opposition. Indeed it made them all the more suspicious; they thought phrenologists were trying to use the Scriptures as a smoke-screen. In any event, the task of Christian Phrenology — to prove that not all materialism was evil and that psychology need not contradict Christian teachings and morality — this task was undermined by George Combe and a select number of his friends, whose phrenology was clearly going in a more dangerous direction.

3

The fork in phrenology'a path came, more or less, with the *Constitution of Man*. It was a successful book in that it brought (as Combe hoped it would) the message of phrenology to hundreds of thousands of readers. But its vast circulation is important to us here for another reason. The *Constitution of Man* was more than a compendium of natural science and amateur psychology; it was at the same time a handbook of natural religion. The more widely it was read by ordinary people, the more extensive its secular gospel became. In effect it helped to popularise in the nineteenth century the very attitudes which churchmen in the eighteenth had hoped to confine to relatively small intellectual circles. Combe's development of phrenology in the *Constitution of Man* as a new basis of morality thus represented an unmistakable challenge to supernatural religion of any sort.

Before preparing the 'Essay on Human Responsibility' for his friends in Edinburgh, Combe had not given much thought to the purpose of religion. He does not appear to have spent an unusual amount of time thinking about religious issues or reading religious material. His work on that essay in 1826 and later on the *Constitution of Man* helped to crystalise his position. The net result was not a little frightening, and it was several years before Combe had learned to live with his doubts. In the meantime he continued much as before, reassuring friends and critics alike and trying to avoid 'giving offence to our Christian neighbours'. He never expressly denied that human actions could be linked to a deity, and he tactfully emphasised (in the *Constitution of Man* and elsewhere) that the 'original human condition' was the handiwork of an Omnipotent Being. He even tried to calm his opponents with vague reminders of his own evangelical upbringing and education. 'I still have seats in Dr Gordon's Church', he wrote to one angry anti-phrenologist; 'I am no stranger to the contents of the Bible, and I make it a rule to do no business on Sunday.'[56] Those closer to Combe knew how misleading such assurances were. Eventually his professions of faith, so often repeated in the early 1830s, brought a fraternal rebuke from Dr Andrew Combe. He chided his brother for giving the impression that he was a regular churchgoer, 'when in reality you only had *a seat* in Gordon's church for a year or two, and did not sit'.[57] By the time this correction was made, however, George Combe had come out of hiding. No longer was there any point in apologising for the path which he had obviously decided to take. The course of his career in the 1830s and 1840s had shown that he too had come to the crosssroads with the *Constitution of Man* and that he regarded the scriptural and nonsectarian religion of Christian Phrenology as too old-fashioned.

Combe's progress towards disbelief is fairly well documented. In 1828 — the same year that he claimed to observe the Sabbath and to keep seats in a neighbourhood church — Combe asked himself what the purpose and advantage of supernatural religion might be. The answer, which he confided to his notebook, was admittedly 'the one most commonly given': salvation. Of course Combe would have been the first to concede that his education was responsible for this reply, but he could no longer accept it. With phrenology as his guide Combe had come to believe that religion's purpose and its effects must depend on the mental powers of the individual. The whole concept of salvation in an unknowable afterlife was unsatisfactory for two reasons: it was base and selfish to define salvation as the deliverance from divine wrath or curse; and phrenologically it was impossible to love a person (divine or otherwise) whom one also feared.[58] Having dispensed with salvation and with the divine interference which that implied, Combe went on to discard virtually all tenets of traditional Christianity: the fall of man, the resurrection of the body, the divinity of Christ, and the ability to know anything 'about the person, place of residence, or the mode of existence of the Deity'.[59] From this it followed that any sort of ritual or formal worship was altogether pointless, because God 'requires nothing from us in the form of service to Himself personally.'[60] Only an inscrutable God was left — and just barely. Everything else was swept away by an uncompromising natural religion which Combe eventually shared with Watson, Mackenzie, Ellis, Simpson and with the cautious Richard Cobden.

While this small band of 'honest doubters' did not attach any special name to their variety of disbelief, outsiders occasionally referred to it as 'Combeism'. The expression seems to have originated in America, where Combe encountered it (much to his satisfaction) during his visit in 1840. The name was certainly appropriate. It implied a mixture of Combe's phrenology, as the most practical human psychology ever devised, and the unwavering sort of natural religion which Combe usually confined to his letters. Certainly, too, the elements of the mixture suggested that the 'Combeists' had grown weary of trying to reconcile Old Religion with their new Science. Indeed it is probably more accurate to say that the Combeists were exasperated, rather than weary, by years of debating with opponents like Philip Jones, who argued that phrenology was not only 'injurious to individuals and families', but had been 'tried by the Word of God, and proved to be Anti-Christ'. The Combeists no longer saw any point in trying to meet such opponents half-way by claiming, as the Christian Phrenologists did, a reverence for the Scriptures. They wondered who could possibly succeed in proving anything to men who were so hopelessly narrow-minded, who opposed scientific inquiry, who dragged their feet on national education, and who were (in Simpson's opinion) probably all ultra-Tories into the bargain.[61] But if the debates faded away, the emnity persisted. And the anti-phrenologists, for their part, knew that Combeists like William Ellis remained at heart dangerous men for their religious scepticism and indifference.

William Ellis (1800–81) was in many ways representative of those called Combeists. He represented a special personal challenge to Old Religion because he enjoyed a national reputation as an educator. None of his fellow phrenologists – not even George Combe – had quite as many connections or as much influence throughout the country. Ellis began his career as a man of business; he was so hard-working that he was reputed never to have taken a holiday. He did, however, become a member of the Utilitarian Society and a friend of John Stuart Mill, whose philosophy Ellis decided to apply to education. His extensive lecture tours, during which he proclaimed his ideas for reformed national education, were financially supported by Lord Brougham, who also encouraged Ellis to establish the first Birkbeck school in London in 1848. Ellis surveyed his experiments from the lofty seat of his own self-confidence and he philosophised about the nature of man. The short essay which he published in 1852 under the title *What Am I?* did not carry his name nor did it have to; his clerical antagonists recognised Ellis's 'philosophy of truisms and commonplaces'. And Ellis knew precisely how they would object. To the expected charge that he made no reference to man as a religious being, Ellis replied assuredly, 'I certainly do not' – as though it would have been foolish to have done so. The reasons for his omission were simple enough: the religious role of man was too mighty a task which he preferred to leave to others; no single definition of religion would have satisfied all; besides which, Ellis was always careful not to antagonise 'our Jewish fellow-citizens'.[62]

William Ellis's successful work in education was among the reasons why religious opposition to phrenology continued into the fifties. In 1852 Ellis established five schools at his own expense, and one of them (at Peckham) enrolled almost eight-hundred pupils. Ellis never ran short of help; Mill was there to encourage him at every turn and as vice-president of the London Mechanics' Institution (1851) Ellis enjoyed ready access to hundreds of parents who wanted their children to receive a practical education. Ellis equated practical education with the sciences and with what he called 'social economy'; he was also anxious to bring to the children the 'lessons of Industrial Life'. Lessons on the 'constitution of man' were also important, and although Ellis was reluctant to explain the functions of the faculties in any detail to children, he did approve of phrenology as part of a practical education for adults. Ellis's methods were also suspect because they took into account phrenological principles in the training of youth and because these methods won the warm praise of an article which Combe wrote in 1852 for the *Westminster Review*.[63] Combe had endorsed Ellis's projects from the beginning; his support in 1852 was an eager *imprimatur* of Ellis's ideas. In fact, Ellis's schools, inspired by a blend of his utilitarian and phrenological beliefs, became the working models for other phrenologists interested in education. Not only did they all agree to abandon Greek and Latin; they were equally determined to discontinue the use of the Bible. In the Benthamite tradition they were ready to make short work of all religious instruction, and thus brought on themselves the charge of fostering a godless education.

This, surely, was one of the most formidable aspects of phrenology's irreligion. The clergy well understood that in the hands of Ellis and his friends phrenology was not the harmless entertainment of head-reading, but a re-orientation of youth away from God. The undisguised materialism of phrenology was the prelude to secularism. How could anyone doubt this, when (in later years) well-known secularists like George Jacob Holyoake acknowledged the debt which they owed to the thinking of George Combe?[64] The phrenologists' use of physiology to explain human behaviour came as renewed inspiration to fully fledged atheists and to men anxious to eliminate the religious character of education. Free-thinkers as well as atheists derived encouragement from phrenology, which they advertised in their books and journals. Thus, while the philosophy of Combe and Spurzheim was not explicitly atheistic, it was suspicious by the company it kept. As early as 1819 the Reverend Thomas Rennell described phrenology as the New Hydra, whose head of materialism was almost harmless when compared to the heads of Doubt and Dissension which devoured the minds of the young through a programme of godless education. Thirty years later Ellis had to endure the same criticisms, for his continuing friendship with, and his intellectual debt to, George Combe was a facet of his career which Ellis made no effort to conceal.

When they spoke about religion, Combe and his friends often referred to the need for a 'Second Reformation'. It was a curious expression for them to use because this reformation was to have nothing in common with the first. In fact the outcome was expected to be much more drastic. Combe and Ellis were not looking for a new Luther to challenge popes and prelates; nor were they hoping for a new Calvin, to trim away the fat of liturgy. Above all the reformation which they invisaged meant the recognition of a wholly different relationship between man and the rest of Creation. The problem with Old Religion was its refusal to admit a relationship which the Combeists saw in terms of psychology and natural religion. Combe elaborated when he explained to Cobden that the 'fundamental error' of Christian society had always been to 'seek a basis of religion in the supernatural instead of the natural.' This basis, which Old Religion foolishly defended, made it unduly difficult for any two persons to agree on the proper use and objects of man's religious impulse, that inborn faculty of *veneration*. Moreover, these disputes through the centuries amounted to 'an enormous waste of religious, moral and intellectual power'.[65] And for Combe at least, the best feature of the Second Reformation was the impossibility of a Counter-Reformation. 'The scientific-intellectual movement in society cannot be stopped', he wrote to Captain Maconochie, 'and supernatural religions will proportionately decline.'[66]

The Combeists saw William Ellis's new schools as an important part of the 'scientific-intellectual movement'. They were boldly intended to provide only secular instruction which, according to Combe, embraced the 'laws of the material and mental worlds, and their relations to human well-being'. If Ellis were able to accomplish this much in his schools, he would give a practical reality to the secular vision which Combeists proclaimed and enjoyed. And in doing so, religion would find its proper place at last, adapting to the reality of Nature and man's true condition. The 'Second Reformation' which the Combeists intended was therefore a complete transformation through natural morality. Writing about Ellis's schools in the *Westminster Review*, Combe predicted that

> 'After a few generations shall have enjoyed this improved instruction, modifications in religious faith may be expected to follow; but they will be gradually introduced, will rest on moral and intellectual convictions, and be supported by divine truth drawn from the infallible book of revelation in Nature.'

What Combe was promising, in effect, was that a system of education based on natural morality would rescue the faculty of *veneration* from centuries of abuse:

> 'Under the present system, religious belief is hastening to a state of anarchy. The Bible is undergoing a criticism of reason, such as it was never before exposed to, and the discoveries of science are daily shaking the established interpretations of it to the foundation. To ignore natural science in our common schools is not the way to strengthen the falling faith, but the reverse; it tends to encourage atheism.'[67]

Who could accept the paradox that the 'godless education' favoured by Combe, Ellis and their friends would ultimately save society from atheism? Certainly not the men of Old Religion! They knew that Combeism could lead only to the loss of faith: the best example was Marian Evans, who proofread the drafts of Combe's essays on *Natural Theology* and *Science and Religion*. Not even the Christian Phrenologists were happy with the idea that religious worship was becoming more rational all the time and that religion was a social organism, always in a different state of development. This notion was always central to Combeism. It was raised in the *Constitution of Man* and it derived from Dr Spurzheim, who spoke of 'religious progress' and noted that men had passed from 'various heathenism' to Christianity and were now ready to take the next step.[68]

Such ideas were mere hints of what was to come. The chief discourse of Combeism appeared in 1847 as the ninth chapter of a new edition of the *Constitution of Man*. It was entitled 'On The Relation between Religion and Science' and was later expanded into a separate volume with the title altered to *On the Relation between Science and Religion*. In the essay Combe was not the least apologetic for his views. He openly attacked Old Religion and predicted it would dissipate before the Second Reformation. He declared that traditional Christianity was exhausted and disgraced and that the Bible was a dead letter. In later editions of this work Combe pointed gleefully to the findings of the 1851 census, which suggested that the British were no longer a nation of churchgoers. In the light of this revelation Combe thought religious hostility to phrenology was 'strikingly impolitic and absurd' because dogmatic faith was itself failing to meet the spiritual needs of men.[69] Having said that, Combe applied a bit of salt to the wound. He made it clear that his criticisms applied not to all Christians but only to those who allowed their disputes over baptism, divine grace, the meaning of the Last Supper, and the 'communication of the Holy Spirit' to 'usurp the place of God's revelations in Nature'.

> 'There are indeed liberal sects [wrote Combe] who reject the extreme doctrines of Church standards, and see in Christianity only a religion of love to God and good will to man, and who regard its founder as a sublime instructor, teaching us by precept and example how to live and how to die. To their views of Christianity my objections do not apply.'[70]

Among the 'extreme doctrines' which Combeists rejected out of hand was the belief in an afterlife. Here, too, the Combeists were distinguishable from Christian Phrenologists, who adapted themselves to Old Religion by the vague belief that at some point man would put on incorruption and be fitted for eternity. In the thirties and forties the Combeists took leave of this position. To hold fast to the possibility of afterlife was in their opinion to mistake the real function of *veneration*. There was no evidence in Nature or in the development of the brain to suggest that man was to live forever in any condition; the resurrection of the body rested only on revelation. Seeing was believing, and unlike the transformation of the caterpillar and other changes in Nature, no one had ever seen a man after death — except, of course, those whose faculties of *wonder, imagination* and *ideality* were excessively large.

The two immediate casualties of the denial of afterlife were heaven and hell. If there were a single 'Author of all things', He could not be disposed to vengeance, for vengeance was the invention of the infant Christian mind. The powers and exploits of Satan were foolish and unnatural ideas which stimulated the mind in the wrong direction; Ellis and Mackenzie were annoyed by the reflection that millions of people still regarded Satan as real as the local butcher or the King of France. Hell was doubted for another reason. As part of their reaction to spiritual predestination, the Combeists did not believe that man was so corrupt as to deserve eternal damnation; his moral weakness was not due to the whim of the Creator or to the lingering effect of Adam's mistake, but to the perversion of mental faculties and the refusal to adopt measures of self-restraint. There was no historic 'fall from grace'; instead there were numerous individual rejections of everything that was reasonable and natural. The conclusion of the *Constitution of Man* was repeated more boldly in the *Relation between Science and Religion*: the sins of man were his disobedience and neglect of Nature, and the wages of sin were paid out in this life. Man's concern should not be with the supposed 'fall' of his first parents, but with his own possible degeneration as he indulged himself in 'animal gratifications'.[71] Old Religion's fixation with human depravity was pointless; even if the Combeists admitted for the sake of argument (as Andrew Combe once did) that man might have fallen from 'an original condition', the fact remained that the universe kept its original constitution and that Nature still obeyed the same laws.[72]

From what they had to say about the occurence of miracles, about Holy Writ and the prospects of eternity, and from their approach to Nature 'in the spirit of little children, humble, eager for instruction, and willing to obey',[73] the Combeists might have been dismissed as a new and rather more stringent school of deists. The impression was unavoidable. They rejected Christian doctrines as mere psychic inventions; they insisted that human happiness and morality were to be found in intelligence and hard work, not in prayer and penance;[74] and if they went so far as to concede the existence of the soul, they connected its operation 'to mental and bodily organs'.[75] They made no apologies for their disbelief. They read the new editions of the *Constitution of Man* and delighted in the additional chapter on science and religion, while Christian Phrenologists read their own editions with reassuring chapters on the 'Harmony between Scriptures and Phrenology'. Certainly, too, the impression of a relatively innocent deism was strengthened by later commentaries on Combe and his friends. 'No representation could be farther from the truth', wrote William Jolly (an inspector of schools) in 1877, 'than [the idea] that George Combe was not a religious man, for no part of his nature was stronger than his religious and moral faculties . . . never has there lived a more earnest advocate than he for religious and moral training, as one of the most important elements in the education and progress of mankind.'[76]

Jolly's words were really too kind. In fact the irreligion of the Combeists (particularly in their role as education reformers) went at least as far as contemporary free thought and was not without its own rancour. From the very beginning phrenology contained a sceptical attitude toward the evidence of Christian doctrines and the value of Christianity to society. As early as 1821 Spurzheim asked Combe whether 'religion in general has done more harm or more good to mankind'.[77] The ultimate effect of Combe's career was to bring this very question out of private letters and into the mind of a wider public. Ultimately too, the irreligion of phrenology was to prove so real as to divide the phrenologists themselves. Some of them, like Combe, Ellis, Simpson and Marian Evans, made light of the religious ideas which they no longer accepted; they learned to live with their doubts. In fact it became a feature of Combeism to taunt Old Religion by showing (as Mackenzie did in one essay) just how far those doubts went.[78] Other phrenologists, like William Scott and Richard Whately, remained fascinated by (and even indebted to) phrenology as a psychology, while still wondering whether man was closer to angels than to demons or whether miracles might not be explained after all.

The Combeists of the 1840s and fifties therefore found themselves out of step. They marched neither with the Christian Phrenologists nor with the atheists, and their tempers were never very long when they had to deal with sectarian Christianity of any sort. Convinced that supernatural religion was finished, they removed themselves to the peculiarly narrow and lonely ground between indifference and atheism, where they awaited more tolerant times. They had clearly placed themselves beyond the pale of Christianity, for in rejecting what Mackenzie called 'all the human devices contrived from the teachings of Christ', the Combeists willingly validated the loud alarms of clerics who warned where phrenology might lead. Combeism was lonely ground indeed, not only because it was occupied by a select number of persons but because they insisted that the demise of Old Religion and the freedom of man's *veneration* were both at hand.

Notes

1. Basil Willey, *The Eighteenth-Century Background* (London, 1962), p.133.
2. *Journal*, I, no.2 (1824), pp.233–4.
3. 'Irish Misery' was explained by the misuse of *veneration* in the *Phrenological Journal*, II, no.6 (1824), p.173f. Andrew Combe dealt with the causes of insanity, including 'religious fanaticism', in the *Journal*, VI, no.23 (1830), p.262.
4. Combe to James Stuart, 13 February 1834. NLS 7386/98.
5. Mackenzie to Combe, 16 September 1835. NLS 7235/160.
6. [anon.] *Phrenology* (London [1850?], p.1.
7. Combe to (illeg.), 11 March 1829. NLS 7384/215.
8. Brown to Combe, 22 May 1853. NLS 7331/105–06.
9. Mackenzie to Combe, 19 October 1841. NLS 7261/33.
10. Ellis to Combe, 17 April 1848. NLS 7293/17.
11. Epps, *Horae Phrenologicae*, p.85.
12. Combe to Mann, 31 March 1841. NLS 7388/458.
13. *Journal*, I, no.1 (1823), p.140.
14. *Ibid.*, X, no.52 (1837), pp.600–1.
15. *Constitution of Man* (5th ed.), p.xii.
16. Combe to Simpson, 10 June 1832. NLS 7385/329–30.
17. Carmichael to Combe, 27 December 1820. NLS 7205/18.
18. Watson to Combe, 15 February 1843. NLS 7270/109.
19. Combe, *Elements*, p.176.
20. Combe to Cobden, 29 August 1852. B.M. 43,661/27 (C.P. XV).
21. Combe to (illeg.), 24 April 1828. NLS 7384/90.
22. Eliot to Combe, 22 January 1853. NLS 7333/93.
23. 'By One of Themselves', *An Address to the People of Scotland, occasioned by the present Disputes on the 'Constitution of Man', as These relate to scriptural Instruction, and coercive Provisions for its Diffusion . . .* (Edinburgh, 1836), p.4.
24. Andrew Combe, *Treatise on the physiological and moral*

Management of Infancy . . . (Edinburgh, 1840), p.49
25. Combe, *Relation between Science and Religion*, p.ix.
26. The *Gazette's* article was quoted in the *Diary of John Epps*, p.168.
27. Spurzheim, *View of the Philosophical Principles of Phrenology* (3rd ed.: London, 1845), p.101.
28. *Journal*, I, no.1 (1823), p.128.
29. Rev. Thomas Rennell, *Remarks on Scepticism, especially as it is connected with the Subjects of Organization and Life* . . . (London, 1819), p.101f.
30. *Journal*, XI, no.57 (1838), p.437. Also, Henry Turner, *Phrenology: its Evidences and Inferences*, p.3.
31. Combe, *Elements*, p.170.
32. [anon.] *Phrenology*, p.11.
33. [anon.] *Phrenology and the moral Influences of Phrenology*, p.5.
34. A. C. Grant, 'Combe on Phrenology and Free Will: a Note on Nineteenth-century Secularism', *Journal of the History of Ideas*, XXVI, no.1 (1965), p.144. Ray's article appeared in the *Christian Examiner* for May 1834, pp.221–48.
35. Spurzheim, *View of the Philosophical Principles*, pp.110–13.
36. A. C. Grant, p.143.
37. William Scott, *The Harmony of Phrenology with Scripture: shewn in a Refutation of the philosophical Errors contained in Mr Combe's 'Constitution of Man'* (Edinburgh, 1836).
38. Scott to Combe, n.d. (1826), NLS 7218/70.
39. Watson, *Statistics*, p.11.
40. Richard Church, *Presumptive Evidence of the Truth and Reasonableness of Phrenology* . . . (Chichester, 1833), p.49.
41. A handbill in the Combe Papers, dated 6 October 1835 (NLS 7236/176), announced a series of lectures given by Barlas.
42. Clyne to Combe, 1 September 1823. NLS 7210/45.
43. Goyder to Combe, 17 May 1847. NLS 7285/42.
44. R. J. White, ed., *Political Tracts of Wordsworth, Coleridge and Shelley* (Cambridge, 1953), introduction, p.xxvi.
45. See Joseph Warne's additional chapter to the *Constitution of Man* (Boston, 1835). In preaching, Henry Clarke looked upon the parables of the New Testament as lessons in human psychology given by the Creator of man's body and mind.
46. *Journal*, IX, no.44 (1835), p.335.
47. Unsigned letter to Combe, 10 December 1835. NLS 7236/141.
48. Rev. Joseph Warne, *On the Harmony between the Scriptures and Phrenology* (printed separately from his additional chapter: Edinburgh, 1836), p.10.
49. William Gillespie, *Exposure of the Unchristian and Unphilosophical Principles set forth in Mr George Combe's Work . . . being an Antidote to the Poison of that Publication* (Edinburgh, 1836), p.36.
50. Simpson to Combe, 15 October 1839. NLS 7252/94.
51. Whately to Combe, 27 December 1847. NLS 7284/155.
52. Whately to Combe, 19 May 1832. NLS 7229/119.

53. Whately to Combe, 19 June 1832. NLS 7229/121.
54. Whately, *Elements of Rhetoric* (5th ed.: London, 1836), introduction, p.xiv.
55. *ibid.*, pp.71–2.
56. Combe to Thomas Watson, 22 March 1828. NLS 7377/96.
57. Andrew to George Combe, 12 August 1836. NLS 7238/14.
58. Notebook, 1 September 1828. Combe Papers, NLS.
59. Combe to Simpson, 12 December 1828. NLS 7384/176; Combe to Mackenzie, 3 September 1831. NLS 7385/111.
60. *ibid.*, Combe to Mackenzie.
61. Simpson to Combe, 15 October 1839. NLS 7252/94.
62. [Ellis] *What Am I?*, p.5.
63. *Westminster Review*, n.s. II (July 1852), pp.1–32.
64. G. J. Holyoake, *English Secularism: a Confession of Belief* (Chicago, 1896), p.44; MacCabe, *Life and Letters*, I, p.28.
65. Combe to Cobden, 13 August 1846. NLS 7381/2.
66. Combe to Maconochie, 7 January 1845. NLS 7390/33.
67. *Westminster Review, op.cit*, p.23–4.
68. *Journal*, VI, no.22 (1829), p.176f.
69. *Relation between Science and Religion*, p.xxvi.
70. *ibid.*, p.220.
71. [anon.] *Phrenology*, p.10.
72. Andrew to George Combe, 24 October 1846. NLS 7278/199.
73. *Relation between Science and Religion*, p.191.
74. *ibid.*, p.6.
75. [anon.] *Man: as a physical, moral, religious and intellectual Being, considered phrenologically* (Glasgow, 1844), p.4.
76. Jolly, *George Combe as an Educationalist* (London, 1877), pp.47–48.
77. Spurzheim to Combe, n.d. (1821). NLS 7207/73.
78. Mackenzie, *Three Lectures on the insufficiency of physical Facts for establishing the continued Existence of the Deity; and on the Superiority of the Proofs that may be derived from the Structure of the human Mind, and its Adaption to the external World* (London [1840?]).

Chapter VII   The Remaking of Man

The long confrontation with Old Religion proved to be the most rigorous test of phrenology's endurance as a scientific philosophy. The confrontation was certainly long and often seemed inconclusive. In some debates both phrenology and traditional Christianity were so ably represented that an intelligent spectator might be forgiven if he thought that neither side had clearly won the field. After all, no phrenologist ever denied the Almighty and no clergyman ever insisted on the complete mystery of the brain. But after several decades of such debates it was possible to discern at least one effect. The phrenologists were obliged either to retreat little by little from logical implications of their creed (such as outright materialism), or stand their ground and defend these implications as best they could. In either case it soon became clear that the men who believed in the shape of the head as an index to character also regarded themselves as true social scientists.

And that is what most dismayed the anti-phrenologists. It was easy enough for them to laugh at curious notions of human behaviour, so long as such explanations remained mere ideas. But what if the ideas were applied to the day to day conduct of human affairs, to man's education or to his moral code? If that were attempted — and if it met with any success — the full danger of a scientific secularism would become unavoidable. We have already examined the ideas of phrenology as they related to contemporary philosophy and to medical science; we have seen briefly how those ideas were received in nineteenth-century Britain. Now we want to explore what the anti-phrenologists most feared: the utility of phrenological ideas to society at large. The scope is wide. Phrenologists were to be found in all manner of reform groups. One could profitably look at their work in the Slavery Abolition movement; one could trace their influence in the Temperance Societies; one could even explore the impact of phrenology on popular literature, drama, and in the burgeoning societies devoted to the cause of personal and public health. While our vision of phrenology's usefulness cannot extend across the whole horizon of social activity, at least it will focus on the two concerns which phrenologists themselves thought most important: penal reform and national education. Before examining either of these fields, however, we must understand what the disciples of Dr Spurzheim meant by man's 'capacity for improvement'.

1

Phrenology's first problem in the business of remaking man was the traditional belief in a paradise lost, in a situation where man enjoyed a more perfect state than at present. One of the points at issue between the Combeists and the Scottish Kirk was the question of man's supposed fall from an originally faultless moral character and high intellectual powers. The question involved man's corruption and whether he was more disposed to evil now than he was at the time of his arrival on earth. In more harsh terms, the issue was whether man was condemned to remain brutish or allowed to improve over the course of centuries. The teachings of Old Religion were not optimistic. And even if the phrenologist and anti-phrenologist both accepted the idea of human 'regression' from a better state, they did not attach the same meaning to the word. The one believed that regression (or degeneration) occurred from time to time in individual cases, the natural result of mental imbalance or serious damage to the faculties; the other believed that all mankind suffered from the same spiritual disability which greatly added to the problem of saving souls. If a man did not act morally there were these two explanations for his corruption: either he was influenced by the devious operation of a diseased mind, undoubtably inherited from one of his forebearers, or he was the weak and willing victim of depravity, the moral heritage left him by his parents.

Phrenologically it was unlikely that man was more disposed to evil now than he had been in ancient times.[1] In fact, the people of civilised Europe possessed much stronger moral faculties than did their Roman and barbarian ancestors, whose skulls the phrenologists thought smaller by comparison. Granted the difference might have been slight for so long a period of time, but the historical fact remained that man was improving. His intellectual growth depended on the evolution of his brain and on an increasingly stable and moral society. It was a thoroughly secular explanation. Man was becoming more remarkable for his sense of duty and his powers of perception, and there was nothing supernatural or mysterious about it. Man's moral strength did not depend on his repentence for sins, or on his being left without taxes and corn laws, or even on the Second Coming: it depended on the 'science and truth' of human nature and on the inherent powers of the human intellect.[2] By the same token, human depravity could not be tied to supernatural dogmas, for those dogmas were long preceded by operative natural laws and the advent of man as an intelligent creature.[3]

There is one curious feature about the debate over human improvement. Both traditional religion, which taught that man's condition was no better than his original state, and phrenology, which assumed that man was slowly improving or capable of improvement — both pointed to the existence of a moral elite. One of the arguments of Old Religion was that the salvation of the soul could be known by man's calm and happy adjustment to life in this world, where his success was consistent and visible. Similarly with phrenology. The righteous man was healthy and successful through his obedience to the laws of Nature: his head was visibly noble because his brain comprised 'an equal endowment of every organ and an absolute balance of the temperaments'.[4] Relative perfection was thus possible in both camps and was blessed with a certain material distinction. In either case the elite existed because not every man was moral to the same degree. The point at issue came with phrenology's explanation for moral deficiency.

John Epps gave the standard phrenological answer when he stated that all mental faculties were intrinsically good but liable to abuse which, if uncurtailed, might cause the neglect of moral duty.[5] Immorality, selfishness and the contempt of knowledge were all flaws of the mind, not of the soul. The extreme form of mental depravity was insanity, transmissible from one generation to the next in individual families, or in whole nations of men. All mankind therefore fit into a phrenological triad. At the top level were the elite, the strong moral characters who enjoyed an 'equal endowment' of faculties and who were in effect the makers of civilisation; in the middle was the bulk of humanity, sometimes aware of its condition and of the opportunities for improvement; at the very bottom were those beyond hope, living in phrenology's hell of 'naturally vicious' and irredeemable minds.[6]

There was nothing particularly grim about the picture of society which these divisions composed. Judging by what he saw daily, the phrenologist was certain that (in Europe at least) the first category was many times larger than the third and would become larger still, given the mental evolution of the second category. If men could only be sorted out properly — if the third category could somehow be removed or prevented from contaminating the offspring of the second — then the pace of man's moral progress would quicken. In an article on choosing domestic servants (an important consideration to the middle class, whether they were phrenologists or not), George Combe boasted of phrenology as the useful and long-awaited instrument for separating good men from bad. By the careful examination of heads 'men of good sense would not place viciously disposed individuals in situations where they could injure others, and draw down the vengeance of criminal law on themselves'.[7] This was the very advice British phrenologists hoped to give to the whole nation. If good men were to continue their moral improvement, the methodical inspection of heads was a wise measure, for scattered in the population were a few individuals whose excessively large animal faculties prevented their own progress and endangered that of others. Phrenology contributed to a sorting out process. The objective was not the emergence, at some distant date, of a class of cerebral supermen, because that degree of perfection implied an end to man's evolution as a material creature. Phrenologists had no such end in sight. They doubted that either the human body or the human brain would ever reach a final stage of development. But superior sentiments could now be identified in the mind, and once possessed of self-knowledge the individual might take greater care in checking some impulses and encouraging others. This is what phrenologists meant by the 'capacity for improvement': a sense of control, however slight and varying among persons, over the formation of ideas and attitudes.

While the capacity could be measured with calipers, it was partly inspired by the optimistic notions of the previous century. Both the men of the Enlightenment and of phrenology spoke of the 'formation of character' as a task open to the individual, and both spoke of the 'cultivation of the intellect' as if each person had a particular plot assigned to him alone. The word 'cultivation' was entirely appropriate too, for it was a principle of the Enlightenment, and of phrenology, that conditions of human existence could be so regulated as to encourage the spread of morality; the growth of human intelligence was almost guaranteed if, like plants in a greenhouse, it was subjected to the proper environment. Of course the phrenologists did not imagine that the happiest changes of environment would immediately and irrevocably change the mind, but they were eager (as Sidney Smith pointed out) to give every member of society 'an enducement to educate and improve himself [and] to regulate passions and purify affections'.[8] This is the sort of regulation, the sort of greenhouse, which phrenologists had in mind. If it could be maintained for a couple of centuries, the process would reap dividends in the visible form of a noble head. The perception of time is where phrenologists differed with the social reformers of the previous century and of their own day. They considered the Owenites, for example, truly utopian for expecting the new moral man to appear in only a generation or so. Phrenologists knew that the human brain, while somewhat pliant, could never allow such quick results.

Many who believed in the rational and systematic 'formation of character' were able to follow phrenology up to this point. They accepted the classifications of mind, within the nation and without; they were ready to concede that some minds were more developed than others; they even granted that various departments of the brain might be responsible for different mental attitudes. But they were in more of a hurry. They did not want to believe that immorality was hereditary or unalterable in a person's lifetime. They preferred to believe that the tendency to drunkenness or to a bad temper came with age, conditions of employment, or the wrong sort of friends: moral turpitude did not come at birth. For those anxious to measure the results of moral training in the present generation, and for those who could not accept phrenology's awesome conclusion that 'some men are by Nature wicked',[9] there was the happy alternative of Robert Owen's *New View of Society*.

Owenism must be considered here for its kinship with phrenology and for its claim to the devotion of several important phrenologists. The apparent paradox of dual allegiance is that in the first half of the nineteenth century Owenism served as a different philosophical base for social reform. Its attention to 'external circumstances' seemed to negate all that phrenology said about the inherent and independent development of the brain. The first principle of Owenism was that man's character was formed for, and not by, him. Yet each of the two philosophies was prepared to make room for the other. They were not at cross purposes when one considered their objectives; the *Phrenological Journal* commented that 'both ardently desire to conduct mankind to the greatest possible happiness to which their nature is susceptible'.[10] With a new moral man as their goal, it was not unusual for reformers to drift from one social psychology to the other and then back again, imbibing the language and ideas of Combe and Owen. For men like Charles Bray, E. T. Craig and William Hawkes Smith, the difference between the two psychologies was not insurmountable. In fact their cooperation was often evident. Readers of the *New Moral World* in 1835 discovered backpage advertisements for phrenological books as well as the admission that phrenology was an extremely important aid to the study of man;[11] a few years later readers of the *Phrenological Journal* were treated to articles by William Hawkes

Smith, who declared that the 'fundamental facts of Robert Owen ... on which all the teachings of socialism are based, [are] a complete abstract and condensation of philosophical phrenology.'[12]

Smith's article served an instructive purpose, for most phrenologists were sincerely interested in the mechanics of Owenism. Driven by *conscientiousness* in the quest for moral improvement, they went to see for themselves how Owenite schools and communities were working. As early as 1820 George Combe visited New Lanark and he was pleased with what he saw. The teachers of the infant school at New Lanark had obviously 'studied the dispositions and faculties of the children' whether they identified them phrenologically or not, and it was no wonder that the school prospered.[13] While their explanations of human behaviour may have differed, phrenologists and Owenites often proceeded toward the new moral man in the same way. For one thing, they shared a common philosophical background. They believed it was time for reshaping man, not by the heavy-handed methods of the past but by more gentle and reasonable tactics. They also shared the legacy of secularism: they agreed that it was unfortunate that moral and social reform should ever have got 'mixed up with the question of the credibility of scriptures'[14] — an issue they preferred to ignore. Moreover phrenologists realised that Owenism was 'something more than an economic experiment', and they were gratified to learn that others were also concerned about human psychology, 'investigating the facts of Nature, the constitution of man, and man's true position in the order of Creation'.[15] If socialism purported to explore such things, could the Owenites really be so different?

Robert Owen himself did not think the gap was impossibly wide. He was willing to assign a major role to what he called the 'original constitution of man' which (and here he did not elaborate) processed feelings and convictions independently of the will and stimulated man 'to act and decide his actions'. So far, so good. The problem arose with the admission, by both sides, that man was a 'compound being'. Which aspect of that being exerted the greater force on character — man's 'original constitution' (which included a set of mental faculties) or conditions all around him? 'The only difference between us', wrote Owen to Combe in 1824, 'is that you, with experience, attribute more to nature and less to circumstances; while from extensive experience, I attribute less to nature and more to circumstances. Time will show which of us is right.'[16] Thus everything seemed to depend on the degree to which each side accepted the day to day conditions of life as a factor in the making of character. In a sense this talk of Nature versus external conditions was an academic discussion: the inescapable problem was simply the factor of time. No good phrenologist could ever expect a sudden transformation of character to occur, whatever the work of natural talents and the environment. 'The vessel formed to hold a pint can, by no circumstances short of a change in the original conformation, be made to hold a quart', declared J. D. Holm in response to Owen's confident estimate of the mind's powers.[17] Owen expected too much too soon. His communities would never succeed because they were carelessly populated: they really required men whose moral and intellectual faculties were the 'mainspring of conduct', not men who were always busy arranging and rearranging furniture, footpaths, and gardens.[18] In terms of mental science there were sound reasons why the Owenites, although 'actuated by pure and excellent motives', would never bring about an improved society. They were too impractical and far too idealistic; moreover the phrenologists had personal reasons for their disdain of Owenism, and this was particularly true of George Combe.

Combe was never pleased that his brother Abram (1785–1827) was one of Owenism's most enthusiastic converts. After his excursion to New Lanark with George in 1820, when they were introduced to Owen, Abram Combe experienced a 'great revolution of the mind' which convinced him that 'external circumstances' were more important than cerebral development in the formation of character. Abram, who had begun work as a tanner in Edinburgh in 1807, now devoted most of his time and energy – along with a good deal of money – to several Owenite schemes. Persuaded that Owen's 'new system' had solved the great riddle of moral reformation, Abram became Owen's chief lieutenant in Edinburgh. He established a school similar to the one he had seen at New Lanark, and the popularity of the venture encouraged him to believe that the whole of a large city (such as Edinburgh) could be turned into a cluster of Owenite communities.[19] In the *Metaphorical Sketches of the Old and New Systems* (1823), Abram announced that Owen's ideas for the formation of character were the 'most valuable discovery for human happiness', and many years later Owen returned the compliment by saying that 'no man better understood my views than Abram Combe, and in all his writings the principles I advocate are stated in a clear and forcible manner'.[20]

The compliment came well after Abram's death. Being of fragile health, he never recovered from a state of exhaustion brought on by his strenuous work at the Orbiston community. Embittered by Abram's death, George Combe and other phrenologists became more suspicious of the demands of Owenism and more distrustful of Owen personally. Almost twenty years after his brother's death, George wrote a short biography whose long title, *The Life and dying Testimony of Abram Combe in favour of Robert Owen's New Views*, hinted at the bad feelings which still remained. The biography suggested that in Abram's case socialism was a wistful fancy and a terrible misfortune; it caused him to be foolish and permissive, whereas previously he was properly critical of people and unwilling to lend money.[21] Abram was lost, a victim of the childish dream world of his friends, of the mental delusions and disappointments of a false psychology. In the collapse of the Orbiston experiment following Abram's death, phrenologists saw their fears and warnings fully justified. Such communities, they said, were bound to fail if there was no attempt to select the participants by the quality of their minds. George Combe deplored Owen's supposed indifference about the membership of such communities; he condemned the 'bad and lazy elements' at Orbiston in much the same way as one might condemn the presence of a bad apple in a basket of good fruit. The effect was similar and Orbiston's fate was likely to overtake the communities in America because the individuals chosen were 'too low to act together in intelligence and morality'. At Orbiston the inherent selfishness of such members caused them to appropriate 'much of the joint stock to their own uses and to contribute as little labor as possible' to the support of the community.[22] Phrenologists therefore felt that, thanks to Abram's misfortune, they had the inside story on the failure of the Owenite communities as a means of social reform. Nor was it an exaggeration for phrenologists to claim that they knew what had happened at Orbiston, for on Abram's death many of the papers and records of the community passed to George Combe, who eagerly took 'the most effectual steps', by the sale of 'all growing crops and every movable article', to liquidate the experiment.[23]

Apart from Abram Combe's association with Orbiston, there was another personal incident which put Owenites and phrenologists at odds. Early in 1824 the *Phrenological Journal*, only recently begun, carried a 'Phrenological Analysis of Owen's New Views of Society'. A draft of the article was sent to Owen and returned with his alterations and approval for publication. When the *Journal* appeared, however, he was not happy with the outcome. In a letter to Combe (who had written the article) Owen said that the essay did not fully represent his views and in several instances completely distorted his principles. Owen also charged that the article was 'calculated to do the greatest injury to the cause' by suggesting that Owenism proposed to 'admit the earliest and most unlimited indulgence of *amativeness*.'[24] Owen expressed his dissatisfaction in phrenological language, but the *Journal* was unwilling to make amends. The furore was never really resolved, and the uncommitted reader never knew who was at fault. On George Combe's part, certainly, there was an element of jealousy; ever since his visit to New Lanark four years earlier, he had resented Owen as a rival philosopher-reformer, and he was suspicious of Owen's occasional use of phrenological language. Neither personality was very accommodating; each was quarrelsome, anxious for acclaim, and highly confident of his formulae for the improvement of man. The real heat of their argument derived just as much from personal antagonisms as it did from principles of social psychology.

Although the dispute over the *Journal's* article was fierce while it lasted, there was no reason for permanent animosity between Owenites and phrenologists. The social psychology of Owenism has recently been described as 'incomplete and inconsistent';[25] the same adjectives might well be used with respect to phrenology. The formulae were not so hard and fast that they did not allow for a certain flexibility, and there were many places where the two philosophies overlapped. There were even times when phrenologists and Owenites seemed to be looking over each other's shoulders, borrowing useful language, ideas and examples. In the 1830s Combe wrote that for all its 'essential defects' Owenism had lately absorbed many phrenological ideas, while at the same time Robert Owen (now permitted to write his own article in the *Phrenological Journal*) noted that 'the most intelligent and experienced among the phrenologists have gradually given more importance to the power of external circumstances'.[26] The *New Moral World* extended the olive branch still further when it declared in 1835 that Owenites did admit to 'certain innate and original faculties' which no two persons possessed in exactly the same way but whose 'utility and innate goodness' could not be doubted.[27] Such comments must have made the casual reader think that the arguments between Owenites and phrenologists would surely come to an end, particularly when they read, a few years later, of Dr Andrew Combe's belief in 'parental ascendency' and other external influences effecting the minds and motions of men.[28]

There were, therefore, several broad areas of agreement which minimised the disputes between the two social philosophies. Both phrenologists and Owenites shared an optimistic view of moral improvement and both defined morality in secular terms. Both were interested in controlled experiments which might help to produce the new moral man, or slightly reform the old immoral ones, through education, work, and the removal of temptations. The critical difference at this juncture (at least officially) was that the Owenites thought the change would come quickly, like 'a thief in the night, while phrenologists knew they must wait longer. But, in practice, were they willing to wait? They were no less anxious than other social reformers to get on with the job, and often they were impatient for results — however slight they might be. There were times when even George Combe grew impatient: he once wrote that he felt like 'taking possession of the wretched and the illiterate' and shaking a new life into them.[29] And so the phrenologists were obliged to come to terms with the present and with conditions outside the brain, just as Owen predicted they must. The marvel is not that the phrenologists were eager to test their reform theories and had to adjust their phrenology accordingly, but that they proposed to use some of the rawest material available. Unlike the Owenites, who invited 'good and honest workers' to join their communities, the phrenologists were willing to deal with the depraved and amoral. They did not seek to cure the insane or any other group clearly beyond reform, but they did want to try their theories of improvement on men whom the harsh and narrow spirit of law had confined to prison. It was in this way that phrenologists entered the campaign for the reform of prisons and of criminal law; they believed they could make some impact on the minds of men whom the rest of society regarded as lost.

2

Phrenologists were long acquainted with the state of affairs in British and European prisons. Gall and Spurzheim visited such places to document their theories and to study cases in which inordinate mental faculties led to careers of crime. Many phrenologists thought it within their professional ambit to make routine visits to these forlorn institutions; while on a visit to Dublin in 1829, for example, George Combe visited not only the penitentiary but also the Mercer Hospital and the Richmond Lunatic Asylum and his observations were recorded in the *Phrenological Journal*.[30] To phrenologists these visitations proved the immense harm done by mixing brains of the lost with those of the redeemable; the insane did not belong in general hospitals and prisons, and not every prisoner was a hopelessly criminal case. The greatest failing of such institutions was that they never attempted to treat the mental disposition to crime, which was the far-reaching objective phrenologists set for themselves in the field of penology.

The whole question of crime and punishment seems always to have interested leading phrenologists. Like Hewett Watson they felt that they could profitably revise criminal laws and greatly reduce crime.[31] Andrew Combe's ideas on prison health as a factor in convict rehabilitation appealed to Chadwick; Archbishop Whately tackled prison reform at some length in his *Thoughts on Secondary Punishment*; M. B. Sampson explored the inadequacies of the law in his work on *Criminal Jurisprudence*; and almost every phrenologist was against capital punishment, arguing that public executions badly effected the minds of spectators. Occasionally phrenologists differed on the best practical means for the rehabilitation of men, but their understanding of human nature usually brought them to the same conclusions. They were dealing with men whose mental tendencies were quite apparent. Their case studies in gaols and other institutions led them to believe that most criminals were creatures of impulse, dominated by one or two faculties of excessive size (e.g. *destructiveness* or *combativeness*). As Hewett Watson wrote in the *Statistics* 'Phrenology explains and proves the fact of some individuals being naturally more prone to crime than others.'[32] At the same time, phrenology showed that many criminals were not incurably dangerous; their brains were responsive to moral discipline and training. Whenever a convict's head indicated that the moral and reflective faculties dominated the 'animal propensities', the phrenologist saw signs for hope. He was ready to advise greater trust and benevolence in dealing with such convicts, who, unlike the incorrigible murderer or the insane, were expected to make at least marginal improvement in their own lifetime. What distressed all phrenologists about the penal system in general (and about the Australian settlements in particular) was that proper conditions did not exist for improvement. Much to their dismay, they found that the old schemes of corporal punishment still inspired prison directors. 'Culprits are but perverse and wicked children', declared the *Phrenological Journal*, 'and the more deeply and exclusively you punish and disgrace them, [the more] you harden them, and render them the worse.'[33] If it was ever to become enlightened and scientific, penology must first recognise that no amount of corporal punishment was likely to cure the criminal mind. The best alternative was to turn the penitentiaries into moral hospitals and retraining centres, where those guilty of antisocial behaviour could receive treatment.

Naturally the precise treatment depended on the individual convict.

The Remaking of Man   147

Ideally prison directors occupied a role similar to that of physicians, for criminals had to be approached on an individual basis. What would one think, asked the *Phrenological Journal*, of a medical doctor who treated two patients in the same way, although they suffered from different ailments? 'The case of the mind is parallel', the *Journal* answered decisively, 'and it is only gross ignorance of mental philosophy that can perpetuate the present system of criminal legislation.'[34] Since phrenology was the only valid science of the mind, it alone was able to determine the extent of mental disease and the prospects for treatment. The first prescription was relatively simple. Convicts were to be classified, one by one, and then assigned to groups according to dispositions and intellects. This emphasis on the individual was still, in the 1830s, a fresh concept, but the phrenologists felt particularly obliged to insist upon it. For years their critics had hounded them with the objection that '... if a man is proved by phrenology to have a bad natural character, it is impossible for that man to turn from his evil ways'.[35]

An embarrassing objection, to be sure. The only way out of the dilemma was for phrenologists to agree that unfortunately some men were indeed beyond redemption. There were cases in which the moral and intellectual organs were so deficient that the minds were lost forever.[36] These were the 'Irredeemables', a relatively small number of men who represented the very worst strain of the British race. It was not hard to identify the unfortunates: they could be detected by the shape of their heads, their fixed and desparate expressions, their 'mental bravado' — and a look at their records. It was wise, under the circumstances, not to send these men to Australia. It was better to intern them in special security gaols in Britain, where they might be observed and further classified by competent phrenologists. Whatever their crimes in the past, the irredeemables must remain alive to serve the needs of cerebral inquiry. Of course that still left the redeemable, whose crimes were slight and whose sorrow genuine.

In general terms, rehabilitation meant an increase 'in morality and industry *always*, and in knowledge *very generally*; for vice and ingnorance are usually associated'.[37] While the mechanics of this approach concerned all phrenologists, the details were worked out by George Combe and the German penalogist C. J. A. Mittermaier.[38] In 1843 they published the *Application of Phrenology to Criminal Legislation and Prison Discipline*, much of which appeared in the *Phrenological Journal* in the form of miscellaneous letters. The ideas expressed by Combe and Mittermaier are important, because in many respects they summarised phrenology's advice and experience in the field of penology. Their basic assumptions were shared by phrenologists everywhere. They agreed that convicts required individual attention and mental exercise and that their improvement could only succeed in clean, well-designed buildings, directed by an intelligent and benevolent staff. They also agreed that detention was necessary and must be wholly effective, lest the criminal mind be distracted by the idea of escape. Nor was there ever any doubt that compulsory labour (usually designated by the more polite euphemism of 'Industry') was a necessary ingredient. Now, thanks to phrenology, there was scientific proof that Industry was useful in the correction of criminals. By itself it could not change the mental dispositions; but it could absorb some of the physical energy which the 'animal propensities' might otherwise direct into criminal activity. Work was an 'external stimulus' to the brain. Once a certain task was finished, the mind (and particularly the faculties of *order, constructiveness* and *ideality*) took delight in a job well done and worthy of praise. There were also practical reasons for insisting on Industry as part of the reform process. The faculty of *conscientiousness* gave man a sense of social duty and inspired him to do everything he could to support himself. The present system failed to do this. It did not stimulate a 'sense of duty' among those whom it hoped to restore to society, and at the same time it was wasteful of human energy and funds. 'The secondary punishment of transportation is notoriously costly and unproductive'; the *Phrenological Journal* complained, 'so are the hulks; and so are all the houses of correction and gaols, which do not, by the labour of the inmates, pay the whole or a part of their own expense.'[39] The ideal penal system must be efficient, and it must be cheap.

Here, by their own admission, the phrenologists were indebted to Bentham. A new penal system, involving well-ordered prisons, would undoubtably be expensive, and in the long term the cost was to be met by the 'profitable labour' of the convicts themselves. Even so, no phrenologist ever lost sight of the fact that the real importance of 'Industry' was emotional rather than financial. The problem lay in the timing of the ingredient. Most phrenologists assumed that the convict could be put to work at once, but George Combe disagreed. He argued for an indefinite term of solitary confinement to precede the period of compulsory labour. During his visit to America in 1838–40, Combe saw the measure used by several prisons, and he thought there were good phrenological reasons for its success. Solitary confinement removed all external stimulants, whether good or bad, and eventually brought the brain and the nervous system 'into the highest state of susceptibility for receiving moral and religious impressions'.[40] The desired effect of such confinement, coupled with a 'duly regulated diet', was to restrain the dominant mental faculties as effectively as possible and to allow other faculties, perhaps long disused, to reassert themselves. In time the prisoner would beg for some kind of manual labour to relieve the 'intolerable pain of solitude and idleness' — a request Combe and Mittermaier thought only a matter of time. It was a sign of a very subtle change in the criminal mind; it was a sign that the moral faculties were stirring. It was at this point that a combined programme of industry and moral instruction could begin. Admittedly there was always the possibility that the step was premature; the prisoner might remain scheming and unrepentant. But that did not matter. There was no timetable to keep, and there was no reason to repeat the cycle indefinitely. The reformer was dealing with deeply-rooted attitudes, and he had to appreciate that not every rescue mission was going to be successful. But for those who did respond to the moral instruction, Combe advised a gradual relaxation of seclusion and discipline, although he felt that the close scrutiny of their conduct must continue.

While no one denied the place of 'Industry' in reforming men, there was one fundamental problem. To phrenologists it seemed that there was too heavy an accent on compulsory labour. The emphasis was understandable, perhaps, because society wanted almost desperately to extract some good from convict hands. It was a great mistake. In practice it had led to the neglect of 'anything which [resembled] moral and intellectual exercise',[41] and for this reason the priority had to be changed. The chief ingredient was more properly 'morality'. Just as the criminal body required detention and deprivation, so too the criminal mind required challenge and a consistent routine. Hard labour without a lesson was as pointless as a long story without a moral. In practical terms morality meant regular instruction. The mind was to be subjected constantly to stories and maxims which pointed to the futility and unreasonableness of crime. Even the rational convict would accept his social responsibilities more readily and would more fully appreciate the virtues of hard work if he first understood their relationship to the laws of Nature. Moral instruction therefore followed a rational pattern. Each convict was to be treated to a series of discussions with a qualified counsellor; the meetings would differ in number according to the individual's needs and were likened to private tutorials. First the convict was reminded that, insignificant though he was, he had defied the harmony of Creation. Predictably he had lost. He was also challenged to defend a behaviour which was both anti-social and irrational. The object of these meetings was to prove to the convict-pupil that each person in society, whatever his limitations, had a natural role to play in harmony with Creation. If occasionally this objective was advanced by an appeal to the faculty of *veneration*, so much the better: chaplains could then be impressed into service, as they were in many American prisons, 'to administer the reproofs and consolations of religion'.[42] Generally, however, phrenologists did not care to include ministers of religion in their plans; perhaps they feared that clergymen would only confuse convicts with sectarian ideas. The concepts of natural morality, on the other hand, were brief and concise, and were easily introduced to the mind. Only the incorrigible would refuse to make these concepts his own. 'The intentions of Nature are always simple', wrote M. B. Sampson in his book on criminal jurisprudence. 'They have only to be clearly stated to be understood.'[43]

At first glance many of these proposals appear rather stringent. In particular, Combe's support of solitary confinement may even suggest twentieth-century 'brain-washing' techniques. There were, however, both theoretical and practical differences. The phrenologists never imagined that they could empty the brain of its accumulated experience and its inherent powers. They never proposed to submit convicts to endless interrogation or to brutal punishment of any sort. Corporal punishment merely corrupted those who had to employ it, and no rational programme of human correction should try to bring its subject to a state of physical and mental exhaustion. Pain and suffering had no place in the benevolent new system. Solitary confinement did, however, because it was thought to be one of those few measures which 'diminish the vehemence of the desires which lead to crime'.[44] Besides, Combe himself was not entirely pleased to defend solitary confinement. He believed that only in collective or 'social' internment did the criminal appreciate his responsibilities. If the method envisaged by Combe and his friends appeared more rigorous than the existing system, perhaps it was because they were reacting (as were so many other reformers) to the abuses and to the lack of discipline which they thought all too prevalant in gaols and convict colonies. They also thought that a new penal system needed a strong initial gesture in order to create a 'mental effect' among the convicts and to prove that the government meant business in their discipline and reformation. The penal system which Combe and Mittermaier derived from phrenology was, on balance, no more strigent than the one already in operation; if anything, it was more enlightened. Far from being harsh, it was flexible. One of Combe's suggestions, made in the *Application of Phrenology to Criminal Legislation*, was to locate the new prisons in the countryside near small towns. This would permit the inmates 'regulated communications' with rural folk, thus giving them a taste of the simplicity, frugality, hard work and all the other virtues normally associated with life in the country.[45] It was hardly a rigid approach to the remaking of man. Instead it was a system which supposed that the brain was the principal agent of its own correction. There was no hint of domination, physical or mental, of one man over another. Above all, the object of the new penology was to better man, if that were possible, not to break him.

One of the boldest if least original proposals in phrenology's plan for criminal reform was the abolition of the death penalty. Phrenology's attack on capital punishment is worth noting, for it made sense to many people who otherwise discounted the value of head-reading in matters of moral suasion and improvement. Since it was a combination of man's brain and 'excessive temptations' all around him which brought him to criminal acts, his wrong-doing was to a large extent unavoidable. Retributive punishment for unavoidable acts — and for an unavoidable mental constitution — seemed a complete absurdity. In Hewett Watson's opinion, society had the obligation to protect itself by destroying those few individuals who were totally unable to resist the propensity to serious crime, but in the majority of cases the death penalty was like cutting down a tree and casting it into the fire after it had borne bad fruit.[46] Man's mental constitution was the best single indication of what his moral fruits were likely to be and the truly dangerous men were easily sorted out from those who were only 'moral patients'. In this light the death penalty was an admission of failure and futility; it solved nothing and exasperated public feeling. As a punishment it had never really deterred men from acts such as murder; their animal propensities craved immediate gratification too strongly to be restrained. In Combe's words, the punishment of death was a 'clumsy moral expedient'. Worse still, it had a disturbing effect on society as a whole: it deadened the 'moral responsibility' and the sensitivity of the spectators and 'increased the cruelty of their dispositions'.[47] Society was therefore advised to discontinue the practice and instead to watch more closely those men whose faculties plainly foretold trouble.

Very few phrenologists did not commit themselves to this aspect of criminal reform. They joined the outcry against capital punishment as a means of 'moral pressure'. Their exchange of ideas, in the phrenologist circles of Germany, Denmark, Britain and America, helped them to reach a fairly uniform policy on penal reform and the agreement that capital punishment served no purpose. *The Phrenological Journal* left no doubt on the issue.

> 'We are of those who advocate THE ABOLITION OF CAPITAL PUNISHMENT IN ALL CASES WHATEVER. In offences against property, we deny both the right to inflict capital punishment, and the expediency; and, although in crimes against limb and life we may grant the right, we utterly deny the expediency.'[48]

Phrenologists in Britain were often members of anti-capital punishment societies, and occasionally they became elected officers of such groups, as Combe did in Edinburgh and Elliotson in London. There is evidence that their membership and their work were very much appreciated by such societies, and there is also evidence that phrenological ideas were accepted as solid arguments by many of these reformers. Spencer Hall, an officer of the 'Society for the Abolition of the Punishment of Death' in London, wrote to George Combe in 1847 that he fully approved of the application of phrenology to penal reform; moreover, he knew many of his colleagues felt the same way, although they were reluctant to admit phrenology's contribution.[49] And yet it was not always possible to overlook this contribution. Through the 1850s phrenologists continued to write about penal conditions, and Combe was invited to lend his advice and talents as a writer to a number of journals, particularly on the subject of capital punishment.[50]

The question remains why phrenologists in Britain, like those on the continent, should have won as much attention as they did from others interested in penal reform. Their proposals were hardly novel. Nor was it their visits to the prisons which set them apart from other critics of the penal system. Not even their conclusions were unique. On the contrary, the utilitarian character of their programme was unmistakeable. As William Ellis assured George Combe in 1853, 'There is nothing that you advocate that Jeremy Bentham or James Mill would have hesitated to concur in'.[51] How gratifying to the faculty of *self-esteem*, to be placed in such company! But was it enough merely to share the opinions of great men? Phrenologists thought they did more than that. They believed that they alone had discovered the scientific basis for a new penology; they alone insisted on treating diseased brains. They alone knew that if Mr Owen's 'New Moral World' were ever to come into being, it must include every man capable of mental improvement. Here at last was the most intriguing (and the most radical) feature of the phrenological system: the possibility of weeding the country of inferior minds. No part of the community, however small or hapless, could be neglected. Every man was of individual importance to the phrenologists, and no man could be left out, free to perpetuate animal instincts in the minds of his offspring. Society was obliged to direct greater energy to the correction of the lawless; little was gained by shipping men off to the Antipodes. Halfway around the world, the convicts remained British. They were a British responsibility. Many of them found their way back to the Britain whose laws they had infringed. Did they return with vicious habits renewed or restrained? To the Combes and their friends, it was folly to fill a continent with the diseased brains of desperate men, bereft of proper discipline and care. The practice seemed only to postpone the day of reckoning and to seal the fate of the Pacific empire. So much was at stake, at home and far away, that the Transportation System could not remain what it was.

3

Phrenologically, everything about the Transportation System was wrong. The occasional report which reached Britain from the Phrenological Society of Sydney was almost always depressing.[52] The society was convinced that the government had no objective in the Australian colonies beyond the creation of a community of outcasts. There was no uniform attempt to separate men according to the magnitude of their crimes; everyone was sent to what Archbishop Whately called 'that commonwealth of thieves'. George Combe, in a letter to D. G. Hallyburton, a phrenologist member of parliament, observed that the men sent to New South Wales were 'the most atrocious scoundrels, who are hired out to masters whom they murder'.[53] Phrenologists were understandably worried about the cerebral climate of Australia: if mental improvement was going to take generations in north-western Europe (where already the best minds were to be found), then clearly the challenge of the Antipodes was staggering. In only a few decades a society infamous for moral regression had come into existence. James Simpson noted the unhappy statistics: each year in Australia there was one execution for every 900 inhabitants, whereas in Britain the ratio was one for every 280,000. The system only guaranteed the continuity of violence, for according to

Simpson the annual cost of police security in the Sydney area alone was £20,000.[54] It was not until 1842 that the colonial officers were given 'any comprehensive body of instructions' to guide them in the discipline and treatment of convicts.[55] What was to be done in the meantime? Should violence continue undiminished, with murder commonplace? Was it possible to improve the situation, or was 'police security' only an expensive dream? These were the questions to which, more than ever before, people were demanding answers. For their part, the British phrenologists detected the makings of real tragedy. Free men of good character were going to Australia in greater numbers, and the long-term danger to their minds and morals was obvious.[56] Action had to be taken while there was yet time, and in 1836 a group of British phrenologists decided to petition the government to recognise the wisdom of phrenology as applied to the Transportation System.

The work for the petition began in Newcastle-upon-Tyne. Phrenologists were strong in this part of the country and closely allied to the northern (as opposed to the London) school of phrenology. On this occasion the Newcastle members were inspired by no less distinguished a Scot than Sir George Mackenzie. Visiting Newcastle with George Combe in 1835, Mackenzie suggested to the society that they address the men in London responsible for the state of affairs in Australia. Together they drafted a letter to Lord Glenelg, the Colonial Secretary in Melbourne's second ministry. The letter (for which Mackenzie later took full credit) was circulated among phrenologists everywhere. They were urged to deluge Glenelg with supporting letters of their own. Every man had his task: George Combe, whose approval of the draft was considered essential, was asked to obtain the support of his Irish friends. The project brought a moment of unity to phrenologists of different opinions. In London, for example, John Elliotson, who was about to lead southern phrenologists into the heresy of mesmerism, committed his followers to the cause. North and south, the local societies gave freely of their time and pledged their support to the petition; they held open meetings in mechanics' institutes and in lecture halls.

The real impact of this exercise in public opinion is difficult to gauge. Certainly Mackenzie did not expect the petition to gain as much momentum as it did; nor did the Combes and the London phrenologists. The momentum might well have been greater had the press not ignored the petition. A few local papers thought the suggestions were sensible, but Charles Maclaren, editor of the *Scotsman* in Edinburgh (and himself a phrenologist), was the only important journalist to commit himself to Mackenzie's proposals. At the moment there was no-one else. M. B. Sampson had not yet come to his position at *The Times*; Archibald Prentice was at odds with his friends at the Manchester *Guardian*. Nevertheless the project was a unique revelation of phrenology's influence in society. Signatures included those of educators, gentlemen farmers, prison directors, members of the Royal Irish Academy, the Royal Colleges of Surgeons, two university provosts, the archbishop of Dublin, a number of lawyers, and a couple of radical M.P.s, who declared 'without hesitation' their firm belief that only phrenology could save the Australian colonies from certain moral disaster. A number of physicians and aldermen also addressed Glenelg on their own, warning him that history would not treat him gently if he failed to act on the petition's advice.

The document which the Secretary for the Colonies received early in 1836 was a brief review of the essential maxims of phrenology.[57] Glenelg was reminded that men differed widely in rank and ability and that there was good reason for this disparity. Historically it was, in the words of the petition, a 'mere delusion' to think that men were born without definite mental dispositions, or as so many 'empty slates'. Each person had a peculiar mental constitution which set him apart from everyone else; to send convicts indiscriminately to Australia merely aggravated the problem. The Secretary was asked to recognise that only a 'true philosophy of man' could distinguish which convicts might be saved and which were not worth the effort. None of these principles obliged the government to stop the Transportation System altogether; phrenology's advice was not as simple as that. The better convicts should still be sent to Australia, and the worst kept for discipline at home. Competent phrenologists were to help in the process of selection; presumably it was they who were to staff 'a special department of colonial government' to put the penal settlements in order. Surprisingly the phrenologists had nothing to say about the conditions of the convicts in transit, but they finished their petition with a challenge. Should the government desire proof of their ability to identify different criminal types, they were willing to examine any fifty convicts the Secretary might choose: the judgments were certain to confirm the official records.

Similar proposals had been made before, and with embarrassing results. A group of London phrenologists, led by the popular lecturer James Deville, advised greater caution in the transportation of criminals. Deville, a man of humble origins who kept the largest single collection of plaster casts and skulls in Britain, badgered London society with his ideas of reforming the Transportation System. He convinced a number of people, at least for a while. He was even reputed to have persuaded the Duke of Wellington, who privately asked Deville for a phrenological explanation of various crimes and perversions. Indeed, the *Phrenological Journal* reported that Wellington very probably asked for a phrenological analysis of himself.[58] In 1826 Deville and his followers went down to the docks to examine the heads of convicts and to warn ship captains who were the most dangerous members of their cargo. The phrenologists were endured for a short time, although a few leading physicians later spoke out in their defence.[59] The others made the phrenologists feel distinctly unwelcome, and they retreated, their counsel ignored.

The phrenologists of 1836 suffered no such disgrace, nor did they come any closer to success. In many respects, their petition was an impressive business. The document itself, clear and courteous in style, presented the views of private citizens, men of good rank and local esteem. Only when it became clear that the Secretary was not about to accommodate them did they lose their patience.[60] When eventually he replied, Glenelg made his position all too plain. It was impossible, he said, to accept such proposals, first because of the cost and secondly because he had no faith in them. Moreover he thought that the request should really engage the attention of the Home Department, and thus forwarded all the papers to Lord John Russell. A useful expedient, to be sure; and for Glenelg perhaps a necessary one. He had only recently come to his job, and he was not very fond of it; somehow he had more than his share of political enemies. With Russell the petition got no further. He politely promised to give it his full attention but later declared that the papers were misplaced. The phrenologists had to resign themselves to defeat. They did so, thinking that this was not the most disgraceful (or uncommon) fate to befall a petition; and besides, the next year brought ample consolation. The House of Commons Committee on Transportation appeared, chaired by Sir William Molesworth and filled with members bent on change. And of course there was the plight of Glenelg himself, increasingly distressed by imperial issues and broken, finally, by events in Canada. His departure early in 1839 seemed to clear the way for a general review of colonial problems — or so many thought. But while the British phrenologists were encouraged by this change, they had already shifted their ground. By 1839 they no longer pinned their hopes to the whims of politicians. They now hoped to bypass the caustic and vascillating ministers in London and introduce their prescriptions to a whole colony. They were

The Remaking of Man 157

now more interested in a man who, at almost the same time of Mackenzie's petition, set sail for Van Diemen's Land as the chief assistant to the Lieutenant-Governor, Sir John Franklin.

When he took up his duties in Hobart Town in 1836, Captain Alexander Maconochie was conscious of his two-fold capacity. On the one hand he was to serve Franklin, who was an old friend but not an easy man to please. And on the other, he was anxious to find out for himself what conditions existed in the colony, as he quite definitely hoped to apply his own blend of phrenological and philosophical ideas. His appointment was not a happy one, and it did not last long. Maconochie made a number of mistakes, largely because of his inability to remember that Franklin was the man in charge. In Britain phrenologists applauded Maconochie's reports, which confirmed public suspicions about the management of the colony. Understandably these reports only annoyed Franklin, who disliked both the public scrutiny and Maconochie's ambitions. Maconochie soon found himself without a job. However, he still had his friends in London, and among them Sir William Molesworth declared that, as far as he was concerned, Captain Maconochie was an 'authority' in matters of colonial police and convict discipline.[61] Maconochie thus stood by his statements and awaited a reversal of fortunes.

Normally the appointment as commanding officer of the convict colony on Norfolk Island was not considered a stroke of good luck. And yet it was this job which the almost penniless Maconochie eagerly accepted early in 1840. It was there, on a tiny speck of land 900 miles northeast of Sydney, that Maconochie hoped to enjoy a relatively free hand in testing his ideas of moral rehabilitation. He soon realised that the island was not the best place for experiments of any sort. For one thing, it was much too small to allow the permanent and effective separation of convicts according to their mental dispositions. Moreover, as Maconochie later remarked in his book *Norfolk Island* (1847), conditions on the island were so harsh that the men barely remembered any other existence. For years there had been nothing to read; the very possession of a newspaper, however dated, was 'deemed an unpardonable crime'.[62] No-one ever expected to deal with such raw material. Even so, during his four years as administrator, Captain Maconochie introduced knives and forks to the mess halls, built two classrooms, distributed books, and provided small gardens. He often said that he found the island 'a turbulent hell' and left it a 'well-ordered community'.[63]

How much of this accomplishment did Maconochie attribute to the wisdom of phrenology? It must first be said that while he was clearly interested in phrenology, he rarely spoke of it as such. In none of his books and pamphlets does the word phrenology appear. In his biography of Maconochie, Sir John Barry has written that the captain was reluctant to introduce phrenology 'to any considerable extent into the administration of punishment' and has referred to Maconochie's opinion that 'prisoners should be treated according to their acts, not their tendencies'.[64] But this does not mean that Maconochie was afraid to accept unconventional ideas in the realm of psychology. On the contrary, Barry refers to Maconochie's curious belief that the dark skin of savages was caused by a secretion which could also effect 'the material organ of the mind'; he also admits that Maconochie lectured his fellow passengers on phrenology during their long voyage to Tasmania in 1836.[65] Then Barry says no more. Unfortunately this neglects the exchange of ideas which continued for fifteen years between Maconochie and George Combe and their wide agreement on the ways in which certain mental faculties must be encouraged or suppressed.

Rather than use phrenological terms, Maconochie in later years described his regime on Norfolk Island as one of 'kindness without weakness'. While his critics (notably the Edinburgh *Review*) detected too much kindness in the scheme,[66] the phrenologists believed that Maconochie was going in the right direction. His system was designed to deprive men of their 'mental bravado' — the arrogance and the confidence which came from having survived old-fashioned sentences of torture and confinement. If a convict survived all that once or twice, he was in a good position to encourage contempt and disobedience on the part of other convicts, and the wrong attitude would soon prevail in the colony. Phrenologists thought the solution was to undermine the arrogance simply by removing the outdated forms of punishment and by helping convicts to reflect on their self-interest. This was the purpose of what Captain Maconochie called the 'Mark System of Prison Discipline'. By good behaviour, hard work (much of it voluntary), and cooperation with other men, the convicts could accumulate marks or points; so many thousands of marks enabled the convict to win his freedom. Under such a system, the costs of penal settlements might be kept at a minimum, and fixed or irreducible sentences became unnecessary. Instead the goal of freedom became a personal project, accomplished under conditions designed to restore men as useful members of society. Rehabilitation involved a change of heart and mind.

Maconochie never imagined that these concepts were entirely his own invention. He could, of course, draw upon his own experience as a French prisoner of war; more than that, his fieldwork in Tasmania gave him a certain pre-eminence among reform theorists. At the same time, he was aware of what others (particularly Bentham) had said.[67] Before he arrived in Tasmania, Maconochie was familiar with the argument between Whately (who opposed corporal punishment) and Colonel Arthur, the Lieutenant-Governor (who was widely believed to favour it). Maconochie also had a 'broad and general knowledge' of the theories of Baccaria, Eden, Blackstone and Romilly,[68] although he probably did not realise just how far others had foreshadowed him.[69] The so-called Mark System was truly the amalgam of designs, in which the psychology of Spurzheim and Combe also influenced Maconochie's thinking. He was convinced that a large part of the brain was given to the reception, appreciation and instigation of good or 'benevolent' acts. It is not true that Maconochie completely accepted phrenology: he believed that parts of the brain probably had distinct functions, but he doubted that there were a certain number of such faculties or that one could accurately be marked off from another. He disregarded what he felt to be the speculative side of phrenology, or 'cranial cartography'. He did believe, however, that it was possible to 'cultivate Benevolence in the mind'. This much was sound phrenology, because it implied a programme calculated to appeal to particular instincts and attitudes. In his writings, notably the *Principles of the Mark System* (1845) and *Secondary Punishment* (1846), Maconochie accepted the premise that to attack the criminal mind was to attack the source of crime – a notion he further advanced whenever he spoke of 'stimulating mental exertion' or 'appealing to the feelings'.[70] It was this approach which convinced the phrenologists that Maconochie came closest to representing their view. He may not have expressed himself in phrenological language, but his policy was a conscious attempt to come to terms with the criminal mind and to redirect it (in Watson's words) to 'the actual and accepted practices of non-criminal society'.[71]

Because Maconochie, as a national figure, was disposed to phrenology, the phrenologists themselves expected much of him. Indeed, they expected much more than he was able or prepared to do. He remained longer on Norfolk Island than he did in Tasmania, although both appointments ended rather abruptly; he had neither the time nor the proper conditions in which to prove himself. His successor on Norfolk Island preferred a tougher (if not a harsher) policy and much of Maconochie's work was undone. By late 1844 he was again left to his own devices, unemployed and obliged to defend himself in a number of pamphlets and in the pages of the *Spectator*. As long as there was hope of a new appointment, Maconochie chose his words carefully: defend himself as he might, he said nothing which smacked of phrenology. To most of his phrenologist friends, his motives were understandable. They knew, in the 1840s, that few persons at Westminster took their science seriously; any association of Maconochie's proposals with phrenology would prevent their ever being accepted.[72] They did not ask him to convert the public to phrenology, although they had hoped that a longer and more successful career on Norfolk Island would have the same effect. The only exception was George Combe: he grew impatient. In the years following Maconochie's return to Britain, Combe repeatedly demanded that he publicly explain how each step of his 'Benevolent System' rested upon a phrenological principle. Combe even suggested that he quote from the appropriate pages of the *Constitution of Man* and the *Phrenological Journal*, both already kept in the British Museum. 'Lay aside your expediency', wrote Combe to Maconochie in 1846, 'and honestly expound the basis for your proposed improvements'.[73] Characteristically, Combe did not realise that he was asking too much of a man who saw little scientific or medical value in measuring convict heads and who did not want to jeopardise his chances of another position.

The expedient Maconochie continued his course, undaunted by the strictures of George Combe. It was not a question of being uninfluenced by phrenology and by his exchanges with Combe; it was instead a problem of giving credit where credit was due. Maconochie failed to label several of his ideas as phrenological and he was not unique in this, for many others who had consciously or otherwise absorbed phrenological ideas were guilty of the same omission. And yet the fact remains that Maconochie never ceased to kindle the hopes of phrenologists. Throughout the 1850s, when phrenology had lost much of the favour it once enjoyed in intellectual circles, Maconochie remained the friend of phrenologists, writing to them, confiding in them and often agreeing with what they said. Occasionally their discussion of prison reform was a public affair: the exchange of letters in the *Illustrated London News* could hardly have escaped attention.[74] But from the beginning the relationship was not without its illusions. The phrenologists overestimated Maconochie's allegiance to the theories and jargon of their science. And for his part Maconochie certainly over-rated their influence and ability to help him. He knew of Combe's connections among politicians; he knew that phrenologists had gained access to the royal family; was it too must to expect them to say a word on his behalf? It was a cautious friendship, moulded partly by geographical distance and partly by the hope of mutual success. Essentially it was a private debt which Captain Maconochie owed to the pseudo-science of phrenology, but there can be little doubt of its reality. He enjoyed his role in their dreams, and his pledges set them scheming. Once, in a letter to Combe, Maconochie mused that if only he had the opportunity to establish and direct a penal colony of his own, large and important enough to operate along phrenological lines, he would surely do so. The project, which Maconochie planned for the Chatham Islands east of New Zealand, never materialised.[75] Together Maconochie and the phrenologists had to content themselves with the management of prisons in Britain.

Looking back over the history of their involvement in penal reform, James Simpson asked George Combe in 1850 whether phrenology had played any role in the reshaping of criminal legilsation or in changing the Transportation System. Simpson could point to no aspect of penology in which their science had been allowed a free hand.[76] Was it because the commercial head-readers had given phrenology such a bad name, as one London surgeon suggested?[77] Or was it because so much of the phrenologists' advice amounted to common sense, free of any reference to their science? Admittedly it was the second objection which carried the more weight. For once the reader had subtracted the phrenological language from Combe's essays or from M. B. Sampson's published letters, he was left (as Lord Palmerston once confessed) with an interesting and reasonable course of action. Convicts were to be treated as individuals; their improvement was to involve mental coaxing; inveterate criminals were to be segregated from the rest; moral feelings were to be cultivated. But to the phrenologist this advice was more than just common sense. He believed that phrenology showed *why* the 'policies of Benevolence' were practical and *why* the older policies of brutal punishment must be discontinued. Moreover his advice was distinctive for supposing that criminals, like 'noxious plants in a large field', could be identified and removed for the good of the crop. It was ruthless advice, too, for suggesting that all the remaining plants should be subject to a long and careful training — to a 'universal system of preventive education, commenced all but in the cradle, and carried on until useful knowledge and intelligent activity shall improve character and elevate the pursuits'.[78] With these lofty words the *Phrenological Journal* described the cause closest to its heart. It was the cause of 'National Education', rather than penal reform, which provided the opportunity (and some of the acclaim) which phrenologists like James Simpson thought they deserved.

Notes

1. Warne, *On the Harmony between the Scriptures and Phrenology*, p.9.
2. *Journal*, VII, no.28 (1831), p.123.
3. Combe, *Life and Correspondence of Andrew Combe*, p.508.
4. Smith, *Principles of Phrenology*, p.18.
5. Epps, *Horae Phrenologicae*, p.10.
6. *Journal*, VI, no.22 (1829), p.216.
7. *ibid*.
8. Smith, *op.cit.*, p.17.
9. *Journal*, VI, no.22 (1829), p.216.
10. *ibid.*, I, no.2 (1824), p.219.
11. *New Moral World*, no.22 (March 1835), p.170.
12. W. H. Smith, 'Remarks on the Application of Phrenology as a Test of the Practicability of Socialism', in the *Journal*, XIII, no.63 (1840), p.121.
13. George to Andrew Combe, 7 November 1820. NLS 7377/5.
14. W. H. Smith, *op.cit.*, p.120.
15. *ibid.*, p.121.
16. Owen to Combe, 26 February 1824. NLS 7213/157.
17. *Journal*, IX, no.46 (1835), p.492.
18. Combe to Simpson, 10 June 1832. NLS 7385/330.
19. George Combe, *Life and dying Testimony of Abram Combe, in Favour of Robert Owen's New Views of Man and Society* (London, 1844), p.9.
20. *ibid.*, p.3.
21. *ibid.*, p.9.
22. *Journal*, VII, no.28 (1831), p.120.
23. Combe acted as the community's 'law agent': *Life and dying Testimony*, p.24.
24. Owen to Combe, *op. cit.*
25. J. F. C. Harrison, *Quest for the New Moral World* (New York, 1969), p.80f.
26. *Journal*, IX, no.46 (1835), p.490.
27. *New Moral World*, no.24 (April 1835), p.191.
28. Andrew Combe, *Treatise on the physiological and moral Management of Infancy*, p.63f.
29. George to Andrew Combe, 19 December 1841. NLS 7379/44.
30. *Journal*, VI, no.21 (1829), p.80f.
31. Watson, *Statistics*, p.67.
32. *ibid.*, p.66.
33. *Journal*, VII, no.32 (1832), p.501.

34. *ibid.*, VIII, no.35 (1833), p.124.
35. The *Journal* tried to come to terms with this criticism in an early article, I, no.4 (1824), p.558.
36. Watson, *Statistics*, p.66; also George Combe and C. J. A. Mittermaier, *Application of Phrenology to Criminal Legislation and Prison Discipline* (London, 1843), p.17.
37. *Journal*, VII, no.31 (1832), p.387.
38. Mittermaier, a doctor of law at Bonn and Heidelburg, edited the debates of the International Penal and Penitentiary Congress, held at Frankfurt in 1846. He returned to Frankfurt two years later to join the Parliament as a representative of his adopted Baden.
39. *Journal*, VIII, no.39 (1834), p.503.
40. *ibid.*, XVI, no.74 (1843), p.15.
41. *ibid.*, VIII, no.39 (1834), p.488.
42. *ibid.*, p.485.
43. M. B. Sampson, *Criminal Jurisprudence considered in Relation to Mental Organization* (London, 1841), p.5.
44. *Application of Phrenology to Criminal Legislation*, p.9.
45. *ibid.*, pp.15–6.
46. Watson, *Statistics*, p.66.
47. *Journal*, XVI, no.74 (1843), p.4; also Combe to Maconochie, 31 October 1844. NLS 7388/782.
48. *Journal*, VIII, no.39 (1834), p.501.
49. Hall to Combe, 11 December 1847. NLS 7285/73–74.
50. *Illustrated London News*, 5 August 1854, p.114.
51. Ellis to Combe, 30 October 1853. NLS 7333/87.
52. The *Phrenological Journal* (VI, no.22, p.222) refers to the phrenologists in Sydney. The science was probably more popular in Australia in the 1870s.
53. Combe to Hallyburton, 30 January 1836. NLS 7386/476.
54. James Simpson, *The Necessity of Popular Education as a national Object* (London, 1834), p.290.
55. John V. Barry, *Alexander Maconochie of Norfolk Island: a Study of a Pioneer in Penal Reform* (Melbourne, 1958), p.39.
56. During the second quarter of the century, approximately 200,000 free migrants went to Australia, many of them encouraged by policies of assisted passage and landsales.
57. A copy of the petition is in the Combe Papers (NLS 7238/160), dated 25 March 1836. Mackenzie also prepared a pamphlet (*Documents laid before the Rt. Hon. Lord Glenelg; relative to the Convicts sent to New South Wales...*) which included the petition and almost fifty supporting statements.
58. *Journal*, III, no.10 (1826), p.324.
59. Among them was Dr Edward Barlow, a member of the Royal College of Surgeons of Ireland and Fellow of the Royal Medical and Chirurgical Society of London.
60. John Fish, a Newcastle J.P., wrote to Glenelg that he and his legal associates knew that phrenology was correct and that Glenelg 'was therefore wrong in rejecting' their advice. NLS 7238/180.

61. *Hansard* [3rd ser.] LIII (1840), c. 1240f.
62. Alexander Maconochie, *Norfolk Island* (London, 1847), pp.5–6.
63. *ibid.*, p.13.
64. Barry, *op.cit.*, p.124.
65. *ibid.*, p.23.
66. The *Edinburgh Review*, LXXXVI, no.173(1847), p.241. praised Maconochie's intentions but decided he was 'far too lenient and indulgent'.
67. For example, Maconochie's *Comparison between Mr Bentham's Views on Punishment, and Those advocated in connexion with the Mark System* (London, 1847).
68. Barry, *op.cit.*, p.70.
69. The idea of an indefinite or 'task sentence' (as opposed to a time sentence) had long ago been discussed by Dr Benjamin Rush in America and by Whately in an article for the London *Review* in 1829. The suggestion of using 'marks' or points of merit had already been made by two Quakers, G.W. Walker and James Backhouse, whom Maconochie knew.
70 Maconochie, *Norfolk Island*, pp.11–13; *Secondary Punishment* (separate section of *Crime and Punishment*, London, 1846), p.27f.
71. Watson to Combe, 14 March 1848. NLS 7298/42.
72. Simpson to Combe, 17 February 1845. NLS 7277/70.
73. Combe to Maconochie, 24 March 1846. NLS 7390/346.
74. *Illustrated London News*, 12 August 1854, p.127.
75. Maconochie to Combe, 1 and 12 November 1844. NLS 7273/43, 46.
76. Simpson to Combe, 10 May 1850. NLS 7311/20.
77. J.G. Perry to Combe, 3 December 1853. NLS 7335/91.
78. *Journal*, VIII, no.39 (1834), p.504.

Chapter VIII The Philosophy of Education

The treatment which phrenology advised for the diseased minds of criminals still left the vast bulk of mankind unattended. The average citizen also required attention if society were to continue its advance. In 1834 James Simpson noted that seven-eighths of the population were labourers whose standards of morality did little to contribute to the nation's well-being. George Combe was equally worried when he calculated that thirteen or fourteen million people were becoming 'organised machines', whose only happiness in life was the gratification of their animal propensities. The problem was not simply one of numbers. The frightening regime of industrial life was now apparent; the Dundee phrenologists, who helped artisans petition the government about factory conditions in 1832, argued that there was less and less time for exercising the superior mental faculties. So far the damage was slight compared to what it might become in several generations. To protect man from the unwholesome changes in his society, and to insure the strong moral character of the nation, phrenologists advised a course which took into account the compound nature of man. As we have seen from their argument with the Owenites, they believed that human character was formed primarily by the peculiar powers which Nature bestowed on man at birth, and secondly by the various social conditions which surrounded man. In order to improve the human species intellectually, the phrenologist proposed to steer a middle course between these two compounds, always recognising that no mental constitution could be substantially altered but admitting that better institutions – and particularly better schools – might aid the evolution of the mind. It was an optimistic view which equated human progress with a general advance in practical knowledge.

The middle course between Nature and 'external circumstances' was not only a compromise in theory; it was impossible in practice. What the phrenologists hoped to accomplish in better schools was more – decidedly more – than they had any right to expect. Their objective was psychological. 'A good education', wrote George Combe in 1819, 'will repress the manifestation of the lower propensities, and cultivate the superior sentiments.'[1] Years later, James Simpson, speaking in Manchester to the Friends of Education, said much the same thing: the object of education, so far as the individual brain allowed, was the regulation of one's feelings and impulses. But then, predictably, the hopes of the social reformer surpassed the findings of the scientist. Combe went on to declare, quite confidently, that 'education can improve even the worst natures to a considerable extent', and Simpson agreed.[2] There is probably no point in trying to explain their curious optimism — an optimism which, strictly speaking, ran contrary to their science. Suffice it to say that the phrenologists, like so many other social reformers and philanthropists, eventually had to leave some of their theory behind. When confronted by the day to day problems of social experiment, they decided that a comprehensive system of education was the only practical way to create a morally strong society.

Having come to that decision, they were carried away by the conviction that their experiment must succeed. Nothing could stop the progress of the intellect. It created its own momentum and the acceleration was greater now than ever before. The *Phrenological Journal* supposed that

'If the present generation could discover and embrace the extended views of morality and science which their successors a century hence will practically entertain, they could not remain as they now are; they would rapidly advance to the highest point ... attainable.'[3]

The phrenologists stood ready to help man reach that point. Given the importance which they attached to the 'cultivation of the intellect', it is well for us to see what they meant by practical education, and then, in the next chapter, how they hoped to make this education a national reality.

1

To most phrenologists outside the home counties, the *Zoist, a Journal of Cerebral Physiology and Mesmerism* was an unsavoury source of information. It distorted phrenology and it belittled the Combes. But when it declared that the 'progression of the masses' was impossible without a 'training of the intellect', the *Zoist* demonstrated something more than its philosophical connection with orthodox phrenology.[4] For although the followers of Combe and Elliotson may have quarrelled, they realised that education was the most appropriate way of improving society and the most fitting antidote to moral disease.

Phrenologists shared a common philosophy of education. They assumed that the brain was in a very real sense the centre of the learning process and that it was subject to various laws. In his *Philosophy of Necessity*, Charles Bray stated that the 'mind is equally the subject of law with matter'.[5] Two pedagogical principles followed from this assumption. First, the weak or inferior mind was one in which the faculties enjoyed little or no regular exercise, while the alert mind required exercise just as much as the body. Phrenologists could hardly deny that mental exercise depended upon conditions existing outside the brain — that much was proved by the practice of solitary confinement in prison. But the most any external challenge could do was to initiate the complex process of thought, in which various faculties acted on one another in a way peculiar to the individual. This led to the second pedagogical principle: education must not be given in a random fashion. It was a phrenological commandment that teachers be aware of the differences of individual students. Not every child could master the same lesson with the same ease or at the same speed. 'To treat a child as an empty sack' warned one writer, 'and to strive to fill it only with facts will have fatal, degenerating and disasterous consequences.'[6] Phrenology prevented such treatment by enabling the teacher to form 'a correct judgment of individual characters' and to meet his duties accordingly.[7] Underlying both principles, of course, was the important assumption that man was intended to be rational and reflective; by understanding himself and his relationship to the world about him, he must become the master and the conscious observer of his own moral progress.

All these pedagogical ideas were very much in vogue. When Dr John Epps remarked that the animal propensities were located 'behind and under [the skull] and it is education's object to keep them there', he was repeating what H. G. Macnab had said in 1819 about the role of education in controlling 'habits, appetites and animal passions'.[8] When in his *General Observations on the Principles of Education* Sir George Mackenzie demanded an appreciation of 'the nature of man', he was making the same request as numerous rationalists before him. And when they spoke of a new moral man as the product of generations of practical education, phrenologists only anticipated the same outcome as many of their enlightened contemporaries. Phrenologists were traditional, too, in their social expectations of education. Morality was an uplifting of the spirit and did not automatically involve a higher station in life. Moral men did not always seek entry to the world of wealth and power; they accepted their occupations in life and made no trouble.

Just as Malthus hoped that artisans could be helped to understand their social duties and their 'true interests', so did George Combe in the *Essays on Phrenology* assume that universal education would transform only personal conduct. 'Let no one apprehend', Combe wrote, 'that by education we will be in danger of rendering the mass of the lower population disgusted with their employments, and lead them to aspire to elevated destinies. The effects of education are always bounded by the natural capacity of the mind; and Nature has taken care to provide a sufficient supply of men for every rank of life.'[9]

The purpose of education was therefore to demonstrate that personal morality (or the recognition and acceptance of the natural laws), and not personal gain (the neglect of those laws), was man's true concern. The objective was socially universal. Labourers were not to strike or resort to sabotage; manufacturers were not to profit from the poor. Among the phrenologists, as among other social reformers, there existed a great faith in literacy as a guarantee of a law-abiding public; they pointed to G. R. Porter's account of crime-free Iceland, where capital punishment was no longer used, thanks presumably to a careful system of education.[10] It now occurred to phrenologists, as it did to others, that the most dangerous elements in society were among the most poorly educated. Convicts were the obvious example. Some thought that convicts were literate if they could recite the Lord's prayer;[11] but phrenologists thought this was a most unsatisfactory measure of anyone's education. They wanted to see natural morality become so pervasive that all men would express themselves without recourse to violence and dishonesty. The best course was to give people, while still quite young, a limited but sensible training based on a 'few sound ideas' which might encourage them to make their way properly in the world, without extravagance, argument and foolish behaviour.

The emphasis on quiet, resourceful and self-reliant people suggests that phrenologists were addressing themselves to the problem of working-class education. They were relatively unconcerned with reforms in the universities or in the private schools. Whenever and wherever possible, phrenologists established schools for the children of the artisan class, such as the Williams Secular School which George Combe promoted in Edinburgh. The working classes were so numerous that phrenologically it was utter nonsense to imagine a high level of intelligence and morality in the country without their being part of the experiment. At the same time, while sympathy for working-class education was a recognition of the needs of the majority, it was also born of fear. In reviewing the instructive lessons of 1848, phrenologists generally agreed that education had to precede the franchise, lest the masses were manipulated, as in France and Italy, by unscrupulous persons.[12] To leading phrenologists the sequence was logical. As professional men and as members of the middle class, they were inclined to see social harmony as the result of education rather than of higher wages or political power. It was pointless to search for the answer anywhere else. As Combe wrote to Ellis in 1848, the attempts to 'organise labour according to socialism, Fourierism, Owenism or any other "ism" ' were bound to 'end in disappointment' because they did not first improve the mind and train the moral sentiments.[13]

There was a distinctly utilitarian character to the education which phrenologists proposed. It is significant that they frequently resorted to the expression 'practical education', to which they gave additional meaning. Now that it was possible to train a person according to his aptitudes, much of the time and energy previously wasted could be put to better use. No longer was there any point, for instance, in compelling students to compose little pieces of music 'in the style of Rossini' when their faculty of *tune* could barely distinguish between sharps and flats.[14] Nor was it advisable any longer to introduce high school students to metaphysical issues, since the faculties denied man (of any age) the power to discover 'as a matter of direct perception, the beginning or the end, or the essence of anything under the sun'.[15] Education was meant to help the individual enjoy his work and do it well. When James Simpson said, as he often did when addressing audiences of working men, that 'education ought to be equally good to all ranks', he remarked that there was no degradation in labour and that any knowledge which helped man in his toil for food was his natural right.[16] Only the inborn talents of the individual prescribed his course of study. No subject, however practical and modern, was necessarily profitable to everyone. Even drawing and mathematics, which most reformers considered useful, came under close scrutiny from the phrenologists. School subjects were to be used by an active mind in designing roads and bridges and ships, and not to serve as mere recreations of a brain relatively weak in *comparison, size, number* and *imagination*. Phrenologists were particularly worried about the teaching of foreign languages. They found that surprisingly few persons had a strong endowment of the organ of *language*. In their opinion it was far better to have no formal education at all than to endure one geared to the study of ancient tongues. George Combe was unequivocal on the subject: in his *Lectures on Public Education* he said that the stable boy who remained at home was much better off than his friends who may have had the opportunity of going to school. In time the stable boy knew how to rear, select and train horses, while the others could only learn the word for horse in other languages.[17] Of all the subjects traditionally taught, the phrenologists saw the least value in the study of Greek and Latin, and nowhere was their pragmatism so severe as in the rejection of classical education.

What phrenologists disliked most about classical education was its apparent monopoly in the training of the young. In an age of science and commerce it seemed a monstrous error to teach children little more than Latin and Greek – particularly when their minds were not ready for such training. Almost any system of education would be better, said James Simpson, if it abolished the 'exclusiveness of the dead languages and allotted them their proper place as subjects of study'.[18] In 1831 the *Phrenological Journal* exposed the 'classical monopoly' in an article which listed essay topics in the examination of fourteen-year olds at the Edinburgh High School. Among the questions which the *Journal* published were these:—

'No. 1. Was the attack of Saguntum by Hannibal, and the invasion of Italy, justifiable on the reasons which he alleges?
No.10. Give an account of the Athenian and Roman calendars.
No.13. Dicuss the art of augury in ancient times.
No.15. Have poets of philosophers rendered more service to mankind?'[19]

Admittedly the Edinburgh High School had a reputation for its classical education but, as Simpson complained, the study of Greek and Latin was still too much a part of the general picture. And so were all the arguments in its defence. There were hundreds of private tutors who, like Mr Stelling in the *Mill on the Floss*, believed that an exposure to the classics and a bit of geometry was the only way of cultivating the young mind and preparing it for subsequent crops. Whether at the hands of tutors or teachers in a large classroom, the old monopoly of the classics was inescapable. One of Combe's correspondents noted that in recent years a few schools (including, incidentally, the Edinburgh High School) had agreed to the introduction of scientific subjects, but these had never been interwoven into the total fabric of education, which the classics still dominated. The result lacked cohesion and purpose, lamented Doctor Leonard Schmitz, a former rector of the High School, and the occasional assignments in practical knowledge were like 'small patches on an old worn coat'.[20]

Publicly phrenologists were very careful to insist that all they wanted was to give the classics their 'proper place'. Like Combe, Ellis and Simpson professed to be satisfied when Latin and Greek were reduced to the status of two subjects on a level with other subjects in a general tuition system, such as that of the Edinburgh College.[21] Privately, phrenologists were rather more like Bentham, hopeful that someday Greek and Latin would disappear from the curriculum altogether. Their vision of the successful school was not unlike Bentham's Chrestomathic model, where there was little room (and probably less patience) for 'useless studies', and where the real emphasis was on botany, geography, mechanics, and zoology. While their vision was similar, however, the phrenologists did not have to borrow from the utilitarians to prove the failure of classical education; nor did they quote Gibbon or Smith, or retell Locke's story of the Chinese visitor who imagined that all English gentlemen were designed to be professors of dead languages.[22] There were new and more powerful reasons why the study of the classics must be given a 'proper place'.

As the phrenologists understood the mind, a certain degree of growth was possible in each individual if only the brain as a whole received regular exercise. In the process of thought each faculty had a part to play; but if any went unchallenged or unused, there was, even in a sound mind, a risk of imbalance. This was the danger posed by years of classical education. The study of Greek and Latin excited the faculty of *language* and left most of the other intellectual organs inactive. There was, to be sure, no single study which exercised all the faculties at once, but at least the sciences (such as mathematics) engaged the powers of *locality, form, comparison*, and *number*.[23] Moreover, the literature of the classics instilled in the formative mind what Simpson called a 'false morality'. Never was there a more 'sanguinary race than the admired Romans', whose feeble and corrupt minds gratified their thirst for blood, their dishonesty, and their gross sensuality. In short, the ancients were beasts, and classical literature carried the germ of mental relapse and corruption through the ages to the modern reader.[24] Far from ennobling man, an education in the classics tended to make him slightly more dishonest and immoral, superstitious and aggressive.

In his autobiography (which he brought up only to his conversion to phrenology). George Combe recalled the revulsion he felt at being made to study Caesar's *Commentaries*. While the other boys identified themselves with the *nostri*, or Roman soliders, Combe asked what right they had to invade Gaul and Britain: his *nostri* were the people of those lands who fought for their freedom. It seemed to Combe that classical education had always neglected to draw the real moral principle from such literature as the Commentaries; instead the danger of such study, unaccompanied by moral commentary, encouraged in boys the growth of *combativeness* and 'sowed the seeds of Toryism in the yet undeveloped mind'.[25] Did Combe's conclusion not contradict phrenology? Was the impact of classical literature so great as to transform the mind? Phrenologists sidestepped the question in some embarrassment but they could hardly overlook the lasting effects of such instruction. The longer it continued, the greater chance there was of its warping students' attitudes and outlook on life. The peril continued into the universities, where the phrenologists were distressed to find Prince Albert advising the study of 'Northern Mythology' and suggesting essay contests on such topics as 'The Death of Baldur' and 'The Deeds of Thor'. How utterly impractical and unwholesome to the formation of young intellects! Phrenologists thought that Prince Albert should have known better than to make such proposals, because he came from a country where innovations in the classroom were of great value to the whole world.

It is not surprising that phrenologists looked to Germany for inspiration in educational reform. The Germans had the 'highest moral and intellectual endowment in Europe' and their heads were invariably noble in appearance.[26] British phrenologists felt at home in Germany; Cobden and Combe agreed that if ever they had to leave Britain, they would settle in Germany. The mental vigour of the Germans destined them, in spite of all their current troubles, to become the leaders of Europe.[27] Phrenologists often toured Germany, delighting in the company of so many great minds; the Combes made several trips, and so eager was George Combe to oblige his audiences in Frankfurt and Mainz that he began the study of German at the age of fifty-four. In view of the German origins of their own science and the strength of the Teutonic mind, it was naturally of interest to phrenologists to see how these moral and resourceful people were tackling the problems of education. Already the Germans had a rich tradition of experiment; Basedow's school at Dessau, for example, with its substitution of natural religion for the catechism, reason for memory, a 'mark system' of discipline for punishment, and physical education in place of Greek, seemed to many phrenologists the prototype of the ideal school.

From the German model, British phrenologists borrowed a number of ideas which were not inherent to the science of Gall and Spurzheim.

The Philosophy of Education 173

One such feature, which Combe listed in his *Lectures on Popular Education*, was administrative. For the sake of uniformity across the country (and for an improved curricula to do the most good), school districts and buildings must be geared closely to the existing patterns of population. Each parish must support a school, each town a high school, each district a gymnasium, and each province (or group of English counties) a university. Education was to be compulsory between the ages of seven and fourteen; it was to be enforced by penalties (as in Prussia) because the phrenologists did not expect all parents to comply voluntarily.[28] Compulsory education would involve the first two stages only – the gymnasium the phrenologists meant to reserve for further secondary school studies and for the classics. The obvious advantage of the system was its comprehension: accommodation was made for all children without exception; at a most crucial point in his mental development no person would be deprived of seven years of a highly moral and thoroughly practical education.

It was an appealing system, both for its comprehension and simplicity, but in some respects a preferable model was to be found in New England. While the schools of Massachusetts were more distant and less visited than those of the Rhineland, British phrenologists were nonetheless well acquainted with them, thanks largely to the correspondence of Horace Mann, the American phrenologist and friend of the Combes. Mann (1796–1859) had the official capacity which British phrenologists dreamed of having. He was for many years the Secretary of the Massachusetts Board of Education and presided (with some satisfaction) over the relative decline of the private or religious schools in the state. His success was one reason why the British phrenologists so revered Mann: from the chaos of religious quarrels he had apparently brought order. What had come to pass in Massachusetts might yet be accomplished in Britain. Sectarian instruction made a code of natural morality almost impossible. Moreover, there was reason to believe that religious schools were not even doing what they set out to accomplish: Horace Mann found that the 'preceptive parts of the gospel' were less understood by students in private schools than in the Massachusetts state schools.[29] Phrenologists, weary of sectarian controversy in Britain, wanted nothing less than what had already proved feasible in America. While Simpson was careful to leave room for alterations in the British scheme, the system of education which he envisaged in the *Philisophy of Education* was essentially that of the United States.[30]

Like other social reformers, the phrenologists spent much time searching for and describing models which approximated their own plans. And, like other theorists, they hoped that an improved system of education would eventually bring about a profound change in moral character. Already they saw evidence of such a transformation in America and in Germany, where, in the words of one of Combe's essays, the change amounted to a 'new reformation'. Phrenologists knew a similar reformation in Britain must entail a national, compulsory, free and secular system of education, but it seemed both pointless and dangerous to attempt the process in existing schools. Again, like other reformers, the phrenologists found that the time had come to build their own models. The obvious answer was to establish new schools in which a course of practical knowledge met phrenological standards.

2

In the *Philosophy of Education*, James Simpson conceded that 'schools of real knowledge' existed in Great Britain and he listed a number of them.[31] Simpson felt that some schools were successful because, whether the teachers knew it or not, they were challenging the mental faculties in the right way and at the proper time. As the *Phrenological Journal* noted in 1824, some faculties matured earlier than others: *form, comparison* and *language* all developed before *amativeness* and *veneration*. The unfolding of mental powers meant that there were at least three, and perhaps four, different stages of intellectual growth, and that there were studies proper to each stage. As a general rule, geography, grammar, some history and 'general information' could be started early; philosophy and religion were not profitably started until the age of eighteen — long after the compulsory period of education was finished.[32] Applying the principles of mental growth and incorporating the experience of the model schools which already existed, Simpson outlined a comprehensive system of education.

The first phase took place in the infant schools, designed for the ages two to six. Phrenologists made no claim to having invented such schools; they acknowledged the work of Fellenberg and Owen. But they insisted that the infant school have a moral purpose. It was not enough to 'give children innocent amusement while their mothers worked at the mills' (as phrenologists imagined the school at New Lanark to do). Phrenologically it was possible — indeed, necessary — to do more than this. A system of infant schools was the only opportunity to undertake moral training before it was too late, for by the age of six it was exceedingly difficult to 'educate the feelings and improve the dispositions' — which phrenologists thought the objective of infant schools.[33] By the mid-thirties, when Simpson was writing, there were a number of such institutions scattered around the country; directly or indirectly most of them were Owenite in inspiration, but a few were more closely identified with the name Samuel Wilderspin.[34]

The Combes first met Wilderspin in September 1828 and at their request he came to Edinburgh two months later. George Combe arranged for his free use of the Clyde Street Hall and introduced him to the staff of the *Scotsman*. Within a short time, a committee was set up to prepare for what was later called the 'Edinburgh Model Infant School'. The committee, which included Simpson and other Edinburgh phrenologists, found that Wilderspin's methods were entirely agreeable. Combe was not publicly associated with the new school because, as he explained to Spurzheim, he had just finished his 'infidel book' (the *Constitution of Man*) and people were likely to view the whole project as 'an infidel and phrenological job'.[35] Whether or not the school was ungodly and phrenological, there is no doubt about the precedents for its methods. Wilderspin (1792–1866) had already served as master at the Spitalfields Infant School, established in 1820 by friends of Brougham and Mill, and he had already brought his pedagogical ideas before the public in a work entitled *On the Importance of educating Infant Children of the Poor* (1823). Wilderspin borrowed freely from the practices of New Lanark. He was certainly no pioneer in education; Brougham once called him a 'good missionary' because he popularised ideas which he took from others. How then did Wilderspin's infant schools differ from others and why did they serve as the model for the first phase of a phrenological education?

The *Phrenological Journal* came to grips with this question in 1831 when it asserted that it was time to 'show our readers why we are so pleased with Mr Wilderspin's system'. The *Journal* linked Wilderspin's plan with the 'organic laws of Nature'; the classrooms were well-ventilated and immaculately clean; there was a fine open-air playground and the masters had at their disposal 'a hundred devices to keep up the spirits and vivacity' of their pupils. There was no caning and no 'shaming' or disgracing students; the system followed a simple curriculum and used colourful pictures and charts. So far, so good; but in every instance these measures were identical to those at New Lanark and other places. In fact, Wilderspin spent much of his time trying to convince the general public (and the Owenites) that his schools represented an improvement over infant schools of the past.[36] The phrenologists themselves detected a crucial and welcome distinction. Under Wilderspin's system, the classroom and the playground became centres of moral training. Whatever the pupil's location, at work or play, he was to be guided in conversation, rebuked for selfishness, chastised for rough behaviour, and always encouraged to be polite. The supervision was immediate and vigilant. Teachers were to earn their wages as conscious makers of little adults. If it did nothing else, the infant school was meant to 'regulate the passions' and 'stimulate the right dispositions'; this is what Simpson, after reading Locke and other rationalists, thought the 'main point' of any training at that age.[37]

The priority went to exciting *benevolence*, and Wilderspin (in spite of his annoyance for not having been recognised as the true originator of infant schools) seemed to bring the right approach to the whole business. He did not try to separate moral principles from religion, as the Owenites were supposed to do; in this respect his plan won full marks from phrenologists who did not want to suffer the same charge of 'godless education' as did the Owenites and utilitarians. Wilderspin's ethics were simple and practically beyond reproach. His Christianity was so nondenominational that Simpson could forsee no objection from any quarter. To encourage good behaviour and cooperation among the pupils, Wilderspin relied on the lessons of the Bible, and his tactics were careful. Instead of reading from Holy Writ, the teacher used coloured pictures and told the parables in his own words.[38] While completely noncommital in sectarian terms, his method won the approval of many clergymen who were otherwise suspicious of the infant school programme, and phrenologists were delighted to have found a way around religious hostility — at least in the first stage of improved education. Wilderspin's strategy in moral training and his refinement of infant school methods generally satisfied phrenologists that he was the 'father of what are now called Infant Schools, even if he did not invent schools for infants'.[39]

The school in Edinburgh which Wilderspin helped to establish was a respectable institution. Although designed for almost 300 students, it accommodated only a third of that number, but Simpson noted that those who attended were of a 'more respectable class of working people, as the appearance of their clothing suggests'.[40] No doubt the weekly fee of two-pence helped to thin the ranks of the school, and among other things the experiment reminded phrenologists that improved education must be free if it were ever to serve the total population. The Rules of the School stipulated that all pupils had to be vaccinated before admission, and the children were asked to arrive each day with clean face and hands, neat clothing, and short combed hair. Those who were absent three days or late for a week without satisfactory excuse forfeited their right of attendance.[41] It was hoped that the school might attract children of all classes because, in Simpson's estimation, class prejudice should not 'deprive an infant of this mighty blessing.'[42] Notwithstanding his long opposition to phrenology, Lord Advocate Jeffrey regarded the school as a genuine benefit to all of society, and the Catholic bishop of Edinburgh had no scruple in advising his co-religionists to send their children there.[43]

Unfortunately the alliance between Wilderspin and the Edinburgh phrenologists did not last long. They readily agreed to the importance of curbing 'inferior passions' at their earliest appearance and hoped that the infants would 'exercise a steady check upon their less favoured seniors'; they recognised that certain dispositions must be discouraged and that the environment must be orderly, safe and clean.[44] But their agreement wore thin over the qualifications of school masters. Phrenologists imagined that when Wilderspin left Edinburgh to visit other schools and establish new ones, he would appoint only sincere phrenologists as masters. He did no such thing. In Ireland, where he went in 1838 to give lectures and meet with superintendents of infant schools, Wilderspin professed himself a 'semi-convert' to phrenology and declared that phrenologists made good school masters at any level of education, but he saw little point in insisting that his teachers have a background in the science.[45] According to the *Phrenological Journal*, this is precisely where Wilderspin went wrong. His administrative genius could not succeed when his staff 'were not in possession of any scientific knowledge of human nature'. His schools were bound to languish as soon as persons 'deficient in active temperament and energetic qualities' were employed.[46]

Still, it was apparent that the infant school could be adapted to a high purpose. The Manchester Statistical Society found that such schools contrasted admirably with the dame and common schools where 'the teaching is so mechanical'; they were also remarkable for their ability 'to awaken the mental powers and to instill a moral and religious principle'.[47] While the eventual decline of the infant schools was not due to their lack of phrenologist-teachers, it is true that Wilderspin and the Edinburgh phrenologists profited from their association. Phrenology played no small part in helping Wilderspin to develop a psychology of education, a debt he acknowledged in 1835 when, addressing the Select Committee on Education, he compared his system of infant schools with others. Owen's system, he said, tried to teach certain subjects too early; this approach 'injured the brains of young children' and had the effect of making them 'prodigies in infancy and blockheads all the rest of their lives'.[48] Wilderspin, like Captain Maconochie, borrowed theories and methods, often without realising their derivation. What he did understand and accept of phrenology, however, confirmed him in the belief that his goals were sensible.

Apart from his ambivalence towards instructors (which probably came from his own lack of special training), Wilderspin found that phrenologists always supported his infant schools. 'We feel more than slight emotion', announced the *Phrenological Journal*, 'when that idol of phrenology, infant education, is attacked.'[49] A frequent criticism of Wilderspin's schools was that they separated children from their proper guardians; the *Phrenological Journal*, jumping to Wilderspin's defence, observed that (in the case of the Edinburgh infant school) the children were away from home for only six hours each day, during which time the school had the chance to see to the development of *benevolence*. 'Besides', the *Journal* asked, 'will anyone pretend that parents in the lower classes are the best fitted to exercise their children in moral, religious and wholesome habits?'[50] The conclusion was that infant schools were entirely practical, especially if the moral tone of the country was to improve. Far from being just a centre of precocious activity, the infant school was performing a task which many parents could not or would not perform; and, in any case, the education of the feelings had to be well under way before the education of the talents was begun.

At the age of six or seven, the child was to enter the second stage of his education. Phrenologists gave different names to this period which stretched to the age of fourteen; some called it primary school; others, juvenile school. Like the infant school, the second stage was co-educational; phrenologists assumed that boys and girls could 'never be more approvingly together' than in the halls of learning.[51] The study programme which phrenologists devised for the primary school was like a chapter from Priestley's *Course of Liberal Education*. Writing, drawing and arithmetic, all begun in the infant school, were continued, and 'practical lessons in physical objects' (such as glass, paper, glue, flax, leather and cork) were also introduced. In the 1840s and fifties phrenologists established a few schools in which to test the new curriculum (while continuing the moral supervision of the infant school), and of these the most notable was the Williams' Secular School in Edinburgh.

The name of the school gives some hint about its operation. Williams once acknowledged that there was a considerable amount of 'theological odium' attached to the word 'secular', but this did not worry him because the 'religious knowledge' presented in the school 'extends wider and deeper than Christianity has ever pretended to do, and [will] eventually supercede it'.[52] Williams privately defined 'religious knowledge' as something peculiar to improved secular education, a 'purely secular theology, never stepping beyond the limits of this world'.[53] It was, in effect, natural religion, and W. H. Williams was a 'Combeist'. He was not given to compromise in matters of religion. To survive, his school depended on 'local liberality', for Williams would not permit scriptural readings which would qualify his school for government grants. Religious knowledge involved, first and foremost, the human condition – those religious sentiments which, in the words of George Combe and Robert Cox, 'lead to stupendous consequences of good or evil, according as they are well or ill directed'.[54] The Williams School recognised its duty to young impressionable minds: the teachers were told not to meddle with any facet of revealed religion, for true religious knowledge was (in Williams's view) neither preceptive nor doctrinal.[55] Of course this policy was in the long term to arouse religious suspicion and hostility, but Williams always maintained that it was the sectarians themselves who were guilty of 'godless education' by confusing young minds with stories of men commanding the sun and moon to remain still or flying to heaven in chariots of fire.[56]

The founder of the Secular School in Edinburgh was a disciple of William Ellis. William Mattieu Williams (1820–92) first came to Edinburgh to study medicine and, like a number of medical students before him, he became a convinced phrenologist. Like the American Horace Mann, Williams named his eldest son after George Combe. In 1848, the same year that Ellis set up his first Birkbeck School, Williams opened the Secular School. A pamphlet which George Combe forwarded to Cobden described the new school as a 'useful, free and secular institution formed for the working class children of Edinburgh', and among the promoters were Combe and Simpson.[57] The school got off to a good start and fifty children were soon learning the meaning of such words as 'elasticity', 'opaque', 'porous' and 'durable'. One possible reason for the school's success was that Combe and Simpson had learned to stay fairly well in the background while Ellis continually sent his shrewd advice. Williams also imitated Robert Owen's use of public relations at New Lanark by opening the school to visitors on certain days and arranging meetings of staff, students and parents, during which the young children gave an account of themselves. The progress of one such meeting, which Combe reported to Ellis, did much to allay the fears of irreligion among the parents. The children were asked about their duties to one another, to their parents and to the world of creation, and with very little coaxing they cheerfully added that God in his wisdom intended them to obey his laws for their happiness.[58]

The curriculum of the Williams Secular School was not unlike that of the Birkbeck Schools. Lessons in geography, civil history, English composition and political economy were given to everyone, while languages, poetry, music and painting were reserved until later or given to those whose aptitudes clearly fitted them for such study. At about the age of twelve, when the faculties of the mind were ready, the 'elements of mechanical science' were taught; the scholars conducted experiments in gravity, divisibility and inertia. Combe and Simpson stood ready to defend the scientific bias on the grounds that the study of air pumps and optics was more fascinating and less difficult than the study of Greek verbs, but criticisms were surprisingly few. Some, like the veteran anti-phrenologist J. C. Colquhoun, suspected Simpson and Williams of trying to exclude the Bible in favour of the *Constitution of Man*; others, like Stephen Seedair, were afraid that from the pedagogical point of view the 'youthful mind' was being confused by too much utilitarian information.

> 'In all the secular schools we have seen [Seedair reported in his 'Tract for all Time'], children are made acquainted with objects of every description, made to repeat their uses and go through a system of training that might be applied to a parrot, monkey, or other imitative animal.'[59]

While Seedair had the Williams School (among others) in mind, his complaint was not very fair. Phrenologists despised 'rote learning' which they associated with classical education. Moreover, Ellis in his Birbeck schools and Williams in the Secular School were both fond of a rather freewheeling Socratic method, in which discussion was fairly spontaneous. Besides, the practical sciences were only a part of the balanced programme of the Edinburgh Secular School. Williams had always attached great importance to a broad and ever-expanding curricula, revolving about the principal subjects of English, political economy, physical education and scientific knowledge. And by the time Williams started his school in 1848, phrenologists had already contributed much to the teaching of these subjects.

The study of English was of special concern. During his eleven-day testimony before the Select Committee in the summer of 1835, James Simpson was asked whether the study of foreign languages formed part of his system of education. He replied that languages were best taught after the age of fourteen. Necessarily this reserved them for 'ulterior education', or training for a particular profession in life, and it meant that classical and foreign languages did not occur in that period of education which phrenologists would make compulsory. With English it was otherwise. Simpson went on to say that the 'great bulk of the working class need no language but their own' and it was throwing time away to teach them any other.[60] Simpson's opinion was never disputed in the secular schools of Glasgow, Edinburgh and Leith, or, for that matter, in Ellis's Birbeck schools. English expression and English literature were the banners of modern, practical training; occasionally they were raised after great battles. When phrenologists of the 1840s and 1850s talked about the innovation which English represented in so many schools, they invariably thought of a young man who was only twenty-four when, with Simpson, he gave evidence before the Select Committee.

James Dorsay of Glasgow has already appeared in the discussion of phrenology's diffusion. A resourceful but somewhat unruly phrenologist whose devotion to the new science no one ever doubted, Dorsay embarked early upon a teaching career. At the age of sixteen he lectured on phrenology and taught grammar in the Glasgow mechanics' institute. He later moved to the High School of Glasgow where he was the sole instructor in English composition and grammar. Given free rein, Dorsay unleashed an enormous energy. Not only did he teach at the High School: he also offered evening classes in elocution, grammar and composition for adults, and delivered free lectures throughout the year on such subjects as 'the use of logic' and the 'connections of language'.[61] As an educator Dorsay gained recognition quickly and was befriended by Lord Brougham. For several years Dorsay led summer expeditions of British teachers to Germany, where they hoped to be inspired by the latest pedagogical techniques.

As the *Phrenological Journal* noted in 1836, the emergence of separate English departments in schools previously dedicated to classical learning was a rather new phenomenon, and it was a good sign.[62] The supervisors of the Glasgow High School granted Dorsay's request for a completely autonomous department, although they later regretted their decision. Once in control, he could make the innovations he always wanted. As a concession to the classicists, Dorsay taught logic in his department, but subverted it to the needs of composition. Essay topics embraced subjects which students observed daily: the different style of buildings, the utility of certain tools and machines, the origin and use of cargoes unloaded at the docks. The influence of Pestalozzi was at work in Dorsay's classroom as it was in others: grammar and spelling were taught as incidentals to the main activity of writing coherent essays. Dorsay belonged to that generation of teachers who, while they may not have looked upon spelling as a matter of private judgment, wanted to end 'laborious and abortive exercises' — such as spelling.[63] Dorsay improvised, and it was almost impossible to ignore his impact. While the English department of the Glasgow High School was by no means a one man show, Dorsay nonetheless became indispensable. He had helped to keep classical education on the defensive in one of its most famous lairs and, whatever his techniques and personality, his accomplishment in Glasgow was a model for other phrenologists. The expanded services which he instigated for all members of the community, young and old, resulted in a greater diversity of subjects and a boost for a programme of useful knowledge.

As Dorsay was phrenology's champion of English, William Ellis was its spokesman for political economy. In the phrenological schema of improved education which Simpson outlined in 1834, it was noted that the minds of children eleven to fourteen years of age were ready for the cultivation of 'liberal ideas' on the use of money, commerce and manufactures.[64] 'Social Science' was the name Ellis gave to his political economy. It was a comprehensive term, embracing many subjects and designed to fit the youngster into the modern industrial community. As Sir James Clark observed, Ellis was a 'remarkably clear-headed man' because he knew what young minds were capable of learning and helped to prepare them for the ruthless world of commerce.[65] Clark might well have added that the learning process was ruthless, too. Ellis did not compromise with the curriculum of the past. In his schools there was no religious instruction whatever (this alone would have endeared him to Combe); the Bible 'was not even used as a reading book'.[66] The classics were also ignored, as Ellis shared Combe's contempt for 'groping among the rubbish, filth and superstition of by-gone times'.[67] Having destroyed these two standbys of traditional education, Ellis was allowed to fill the school day with the more practical knowledge of chemistry, physiology, mathematics and, above all, political economy. The net result was a rather austere curriculum, almost vocational in nature and consciously utilitarian.

William Ellis was, after all, a utilitarian himself. He acknowledged his philosophical debt to James Mill and valued his friendship with George Grote and John Stuart Mill. But curiously enough, it was his reading of the *Constitution of Man* which brought Ellis into the 'education madness' of his times. The emphasis which Combe placed on 'knowing the world all about us' appealed to him because he thought that the economic life of man was dictated by natural laws and by natural virtues, such as punctuality, honesty and sobriety. In other words, Ellis's 'social science' was a combination of laissez-faire economics and moral discipline. Education, if it were to mean anything at all, must be

> 'in accordance with the constitution of the human mind, and best calculated to strengthen, develop, and rightly direct all its faculties, by presenting to them the objects naturally adapted to call them into varied and healthy activity.'[68]

When Ellis transferred his attentions from business to education in the mid-forties, he discerned a special value in political economy. Properly taught in the schools and mechanics' institutes, it served as a counterweight to social chaos. It reminded people that an understanding of the 'natural economic laws' and the 'free play of economic forces' benefitted everyone, precluding the need for combinations and strikes. This, according to one modern writer, was the soothing message of Richard Whately's *Easy Lessons*.[69] William Ellis and his disciple W. H. Williams, both admirers of Whately, taught the same brand of political economy and for the same reason. They saw human misery not in terms of sweated labour, profiteering and speculation, but in conditions of 'ignorance, unskilfulness [and] depravity of disposition'.[70] Occasionally Ellis and Williams had to contend with the charge of being 'special pleaders for the capitalists', but the accusations of working-class parents did not bother them very much. They knew that if the masses were to be educated political economy was the most appropriate subject. It was the pivot on which all education rested, in the schools which Whately praised and which Ellis and Williams established. When he visited one of Ellis's schools with Combe in 1852, Cobden noted that the curriculum was probably not as varied as in other progressive schools of the day, but at least the teaching of political economy was a fair guarantee of good citizens – particularly from a class whose late behaviour was anything but tranquil. Equally important, Cobden thought that such training prepared children for a more harmonious and cohesive world in the future by instilling a sense of honesty and fair play – the seeds of cooperation and reciprocity at home and abroad.

Among the subjects to which Ellis gave scant attention in his schools was music, a study which, generally speaking, found little favour among phrenologists. The faculty of *tune* was not believed to develop very early in the brain, and it was therefore difficult to plan for extensive musical training at the primary level. While little or no time was spent on introducing and explaining the musical instruments, singing was used as an incidental exercise in most of the schools run by phrenologists. Wilderspin and Williams both provided for singing as an *entr'acte* separating the main subjects of study, and so did Ellis in the Birkbeck schools. And yet, while the opportunities for vocal music were limited, the use of music underlines the secular character of phrenologists' schools. Theoretically songs were useful as an outlet for the emotions and for rousing the mind to fresh activity, but in their schools the phrenologists preferred popular melodies instead of hymns and psalms. Sacred music had its place elsewhere – a point George Combe sought to prove in 1847 when he publicly contributed to a fund for the 'revival of sacred music'. It was a calculated move on his part. He wished to show that phrenologists did not want to eliminate devotional singing, but only put it in its proper place. Combe's gesture was ironic, for the fund was supervised by the Rev. C. J. Kennedy, an old foe of phrenology who suspected Combe of starting the theory of evolution.[71] It was even more ironic, however, that of all those phrenologists who wished to add music to the new curriculum, none was more active than a former German priest, who came to Edinburgh at Combe's invitation in 1841.

Joseph Mainzer (1801–51) was a musician, composer and a musical propagandist. Ordained a priest in 1826, he later abandoned both the religious life and his native Germany. He moved to Brussels in 1833 and there composed an opera, 'Le Triomphe de la Pologne' which apparently was never performed. Still later he moved to Paris, where he became a music critic and composer of dramatic and sacred music. His acceptance in the *salons* of the French capital did not dampen his democratic sentiments. He wrote another opera, 'La Jacquerie', which proved successful and he arranged concerts and song festivals for working-class audiences. According to the *Chamber's Journal*, it was Mainzer's intention to 'bring scientific music within the reach of the humblest as well as the highest classes of society' and it was not long before he had become the 'apostle of popular music'.[72] When one of his concerts attracted 3000 *ouvriers*, however, the government took fright; they imagined that Mainzer was trying to subvert the artisan class. At this point Mainzer decided to pull up stakes once again and move to Britain.

In almost no time Mainzer became a celebrity. He astonished everyone by his quick mastery of English and by his 'aggressive musical journalism'.[73] In Edinburgh Combe, Simpson and Williams were attracted at once to the 'noble and clear-thinking German' and they invited him to support their secular school. Mainzer was happy with the project. He believed that education must have a 'scientific basis' and he had long ago accepted phrenology, although he hesitated to 'apply it too strictly to the nations' because intermarriage had brought about great changes in the old racial characteristics.[74] He was particularly pleased by the nonsectarian (if not secular) nature of education in the phrenologists' schools. Mainzer wanted to see vocal music adapted to the needs of harmonious society; he therefore preferred relatively simple songs which praised various domestic and social virtues such as humility, cooperation, temperance, cleanliness and good order. He shared Combe's opinion that sacred music was best left to somewhat older and specially trained choirs.

Mainzer did not remain in Scotland. He thought that the Scots were either unaware or unappreciative of his talents, and he complained of the 'bigotry and intolerance' which accompanied his attempt to have student choirs sing music by Rossini. On one occasion, in 1847, Mainzer was accused of leading the children into idolatry, and the parents brought him before the directors of the Northern School District. The episode should not suggest inconsistency on Mainzer's part. He had chosen the music for its technical qualities and because he thought the sentiments were acceptable to an 'Episcopal choir'. Moreover, Mainzer probably expected that others approached music as he did: not as expressions of dogma or even of belief but as works of art.

Besides, Mainzer was hoping to get the chair of music at Edinburgh University. As soon as it became clear that his broad and occasionally Catholic tastes in music were against him, he resolved to go to Manchester. Even from that distance, however, Mainzer remained interested in the Edinburgh experiment, and he advised his friends on the proper use of music as an aid to memory. He agreed with them on a relatively minor role for music in the early years of education, during which time it was best made part of a 'relaxing discipline' in the curriculum.[75] Mainzer's contribution was an uncomplicated one: he helped to remold a long-used and religious-oriented exercise into one which more closely complimented a secular pattern. If his work is forgotten today, it is not because of his phrenology or his rather short and mobile career, but the slight importance which education, old and new, has always attached to the study of music.[76]

While vocal music played only an incidental part in the improved curriculum, physical education was among the essential ingredients. In Germany physical education had long ago become part of education — a development which phrenologists thought a reflection of the active German mind. In any case, it was important for education to become a complete process, providing exercise for the physical as well as the mental faculties. Nature intended them all for improvement. In the preface to one of Dr Charles Caldwell's works on the subject, George Combe declared that physical education was the requisite of all successful training, and phrenologists believed that a programme of 'systematic exercises' should begin in the infant schools and continue until the end of formal education.[77] The goal of education was long-range, both physically and morally: the highly complex and noble brain was the product of many generations of cultivation, and so too was the noble body. It was a reasonable assumption if one believed that mind and body each had a physiology of its own which responded to exercise. What phrenologists said about the parallel development of mind and body made sense to many educators and advocates of physical exercise. Samuel Smiles made a point of sending a draft of his first book on *Physical Education* (1838) to the Combes and agreed with them that the physiology of mind and body was an appropriate matter for study in the schools.[78]

Perhaps it is remarkable that of all those who accepted, or who were now willing to accept, physical education as an integral part of the curriculum, only the phrenologists assigned it quite so much importance. Not only did they expect daily exercise for the whole period of compulsory education: they also called for the teaching of physiology and the 'laws of health' (or hygiene). The phrenologists regarded mental and physical development as two inseparable aspects of education, and they never had any patience with those who (like Georges Cabanis in France and James Mill in England) argued that a robust body was actually an impediment to intellectual greatness. In fact Andrew Combe and Robert Macnish thought this a most dangerous view. Was it not the peculiar assembly of brain, physiology and the Four Humours which produced human character? Did an enfeebled body not effect, however slightly, the operation of the mind? And did not the control of one's faculties imply a certain stamina in both mind and limbs? The social science which Ellis taught – the 'science of human well-being' – emphasised the complexity of the human condition: there was a time for exercise and games, for the study of anatomy, and for learning the habits of good health. The same course was prescribed by William Lovett, who like Ellis was inspired by reading the *Constitution of Man* to regard education as a chance to develop all the human faculties as fully as possible.

The ultimate aim of a sound mind and body was almost Malthusian. Judging from their experience in the schools, the phrenologists were certain that ignorance served as a check to the growth of population. There was no doubt that the death-rate would fall if people were more careful about their health and the exercise of their children.[79] Early marriage was another evil; phrenologists warned of unhappy consequences whenever a couple married before the mental faculties had fully developed. On this point the phrenologists were as explicit as Malthus had been: nature was inimical to marriage before the age of twenty to twenty-two for females and twenty-five for males. It was not simply a question of political economy or having time to gain a financial basis for married life; it was a matter of emotional and mental maturity. At one point the *Phrenological Journal* argued the eugenic case in religious terms, stating that before they entered marriage each partner should feel that the 'creator's laws are not violated'; but this hardly disguised their belief in a purely scientific approach to marriage based on the total physiology of the individual.[80] The values which they placed on education – and particularly the education of women – were therefore secular. There was no mention of duty or virtue or strength in the religious sense; the only real meaning of such words lay in the physical and mental well-being of future citizens and parents. It was this conviction which brought Combe, in the *Lectures on Popular Education*, to advise young ladies to prepare for marriage by a programme of physical education and the study of phrenology, an 'indispensible step in the education of young mothers'.[81]

For that matter, the study of phrenology was a step in the education of the young. If he were asked about the purpose of his school, the phrenologist-teacher would have replied that schools existed to help the pupils know themselves. All departments of education were reducible to a variety of self-knowledge. Natural religion was part of this process; so was study of the natural sciences, and an understanding of the senses, anatomy and human reproduction. But even if all this were done, the programme would still be incomplete and old-fashioned. For children to appreciate the unique wonder of the human species and to understand something of man's conduct, it was desirable to introduce phrenology itself into the curriculum. While it had long been possible for adults to attend phrenology lectures at the mechanics' institutes, no special provision was made for the younger audience until the phrenologists opened their own schools in the forties and fifties. In several places phrenology was integrated either with physical education or with the sciences. A printed notice of the Leith Secular School announced in 1854 that the annual examinations would cover the fields of natural philosophy, social economy, and phrenology; at Leith the senior students (aged 11—13) could receive up to two years of instruction in the new mental science.[82] In Glasgow phrenology was taught at the Saint Andrew's Square School and was introduced by James Dorsay as a regular subject (under the name 'mental philosophy') at the High School in 1842.

Here was one reason for the public suspicion which attended the opening of the Williams Secular School in Edinburgh in 1848. Understandably it appeared as though the phrenologists were trying to propagate their science among the young: their schools were a conspiracy designed to insure that the next generation was more receptive to the notions of Spurzheim and Combe. The suspicion was not far from the truth. Combe was sure that phrenology would 'spread like wild-fire' once it was accepted as a science in the schools; from America came encouraging signs that this was already happening. In a letter to Combe, Lucretia Mott observed that while phrenology was not taught in any of the church-affiliated schools of Philadelphia, it was offered in several other places, and she thought that someday phrenology would completely replace the 'dogmas and hidden mysteries' which currently occupied the curriculum.[83] So widespread was the teaching of phrenology in New York and New England schools that the first school edition of the *Constitution of Man* was issued in 1840; Combe offered to send twenty-five copies to Cobden for distribution to various schools in Manchester.[84] How much phrenology the children actually learned was, of course, another matter. Simpson advised against teaching too much too soon, and generally instruction took the relatively easy and pleasant form of 'comparative phrenology', or the cerebral distinctions of different races and cultures. As for the charge that phrenology in primary school was no more useful than the study of Latin, phrenologists believed they had an acceptable answer. At least 'comparative phrenology' was scientific, based on many hundreds of interesting cases and involving anthropological studies of faraway lands; it was a subject calculated to excite the *wonder* of any child by introducing him to the culture and achievements of other people.

With or without phrenology, the new curriculum required a new atmosphere. Practical knowledge was most suitably and effectively offered in places which had no connection with the traditional classrooms. Whenever possible the break with the past was complete. When it was not possible to build new schools of their own, the phrenologists used mechanics' institutes; this was particularly true of the educational services which they provided for women. One of the most frequent visitors at the institutes was Dr William B. Hodgson (1815—80), a thorough-going reformer who as a youth had preached phrenology in the Scottish highlands. In an article for the *Phrenological Journal* in 1841, Hodgson discussed the great role which the mechanics' institutes were expected to have as the proving ground for a system of secular education.[85] Of course the institutes were intended for the use of adults rather than their children, and most of the mechanics' halls were no more suitable, as physical plants, than the old school buildings. Andrew Combe and E. T. Craig, the ventilating engineer, insisted on a happier atmosphere of well-aired rooms, garden courtyards, gymnasia and large playing fields, while the *Phrenological Journal* warned of the cerebral harm which came from taking classes under conditions 'as deplorable as the Black Hole of Calcutta'.[86] Naturally the funding of brand new schools was an expensive business and a real test of local philanthropy, and not a few phrenologists were obliged (as William Ellis was) to convert abandoned chapels and halls to new purposes. This they did in the enthusiastic belief that their schools were only temporary installations. In two or three decades, when the public finally realised the immense value of education reformed by phrenology, the government would commit itself to the phrenological system and then phrenologists would move into impressive new buildings.

The dreams which phrenologists entertained on the subject of new schools, lavish in style and facilities and generously staffed, were not unlike the dreams of Owenites, who hoped to carry principles into practice in the arcadian bliss of landscaped parallelograms. Actually the new atmosphere of learning contained some rather stale elements. The Lancasterian System, for example, was often retained because it was supposed to excite the faculties of *conscientiousness, benevolence* and *veneration* by permitting the students to help teach one another and judge one another's conduct. The phrenological merits of the Lancastèrian System were decisive: it 'encourages the timid, represses the overbearing ... and produces that rational feeling of superiority founded on superior conduct.'[87] The reliance on monitors, who found a place in the phrenologists' infant schools and in the juvenile schools, was regarded as a practical way of teaching, and Ellis and Williams were ready to pay the monitors as much as possible.[88] When George Combe visited Ellis's school in Holborn he was pleased to hear a boy of twelve deliver, entirely on his own, a lesson in Social Economy in 'a distinctively efficient style'. The monitors were also expected to double as 'moral police', helping to insure fair play, cooperation, and *benevolence*.[89] As in Owenite schools, the practice of rewards and place-taking was abolished because, in Combe's words, 'it fosters envy, pride and selfish ambition and [does not] cultivate the love of knowledge for its own sake'. The mind of the student did not always regard the medallion or badge as a certificate of accomplishment, but as a symbol of personal triumph over all the others in his class.[90]

The intense interest which phrenologists showed in the formation of the young mind must not obscure the fact that in the realm of education phrenologists were not pioneers. They were not the first to abolish place-taking and punishments. They approved of the Lancasterian System and did little to modify it. They kept the monitors and attempted to give essentially the same education to boys and girls. They ejected the Bible, as others had done, and sought after German and American examples. The conditions of the new phrenological school house were, like the subjects of the new curriculum, not really novel at all. But the lack of originality does not mean that the phrenologists had nothing to contribute. In the argument which raged between the defenders of classical education and those who advocated radical reforms, the phrenologists closed ranks with the innovators. They were on the side of experiment, practical knowledge and science. No one could really deny that the alliance existed: it delighted some (like Thomas Huxley) while it alarmed others.[91] Writing in 1821, one defender of the grammar schools complained of the growing popularity of the sciences: 'a knowledge of chemistry seems to have become a female accomplishment', wrote Vicesimus Knox, 'and the rising generation of studious youth devote much time to it, as they do to geology, mineralogy and, perhaps, craniology.'[92]

The complaint serves to illustrate merely half of the contribution made by phrenologists to the 'education madness' of their time. Whenever they were allowed — in the mechanics' institutes or in their own schools — they added a new study to the curriculum of useful knowledge. Moreover they helped to make science an eminently practical study, relating it not only to the needs of the classroom, but to the well-being of society generally and to the health of generations to come. In this sense science was not a luxury; it was, properly speaking, the necessary study of life and human conditions. As the Westminster *Review* (not always gentle with phrenologists) asked in 1852, 'Who has not observed how effectively the phrenologists have been inculcating the first principles of physiology, and the art of life which depends on that science, into the public mind?' There was no doubting which phrenologists had done the most to educate the public, for the *Review* continued:

> 'Have not those two public-spirited brothers, George and Andrew Combe, made the reading portion of this whole nation familiar with the idea of the absolute dependence of health on obedience to organic laws? In consequence of the generous labours of these and other benefactors of the same class, how differently do we deal with our infants, how much more do we attend to the ventilation of our dwellings, how much more considerate are we in dietetics and regimen, how much more carefully do we avoid intemperance and irregularity of every kind! . . . '[93]

The other half of their contribution was probably more important, for the phrenologists added much to the clamour on behalf of improved education. While acknowledging the work and the models of others, they were conscious of their mission to explain concepts and methods of education which had been current for some time. Again, even in the eyes of their contemporaries, they were remarkably successful. George Combe never wrote an original or systematic treatise on education, and perhaps he did not have to. His most durable reputation in the nineteenth century was (not unlike Robert Owen) that of a reformer of schools. Few could match his influence. It was his *Moral Philosophy* which brought Barbara Bodichon to her faith in social progress and female emancipation, and it was the *Constitution of Man* which inspired Lovett and William Ellis, who was probably the most influential educator of the mid-century. It was to Combe that people wrote, asking for introductions to Wilderspin and Mainzer or to secure the help of other innovators, and it was Combe who pressed James Simpson to write the *Philosophy of Education*, which embodied all the principles for the remaking of man through education. In effect, it was the role of Combe and his friends to help prove theories of sound elementary education and to transmit those ideas to an ever wider variety of dedicated reformers: to feminists like Barbara Bodichon; to politicians like Sir Thomas Wyse and Richard Cobden; to working men like William Lovett. And although only a few schools owed their existence to the modest philanthropy of phrenologists, it was the popular science of craniology (so contemptible to headmaster Knox in 1821) which helped to rouse the public to the needs of the individual school child.

While it is not strictly accurate to list Combe, Ellis, Simpson, Williams and Hodgson among the educational innovators of the nineteenth century, it is true that they enjoyed a considerable reputation as promotors of experimental schools and sensible courses of study. They knew what direction education must take. At the same time, they did not feel that the reformation was coming quickly enough: their model schools were too few to guarantee the new moral man. How long would the schools survive? Given the religious passions of the day, it was hard to tell; and outside the schools was the land of darkness which phrenologists could never overcome on their own. The solution which eventually occurred to them was not unlike the hope they once cherished to bring about convict reform in Australia. To gain impressive patrons and to convert many more schools to their purposes, the phrenologists would seek friends in high places. With the right connections they might vanquish their sectarian foes from above, and force a new day of knowledge upon every parish in the land.

Notes

1. Combe, *Essays on Phrenology*, p.337.
2. ibid., and *Report of the Speeches delivered at a Dinner given to James Simpson, Esq., by the Friends of Education in Manchester* . . . (Manchester, 1836), p.5.
3. *Journal*, V, no.18 (1828), p.279.
4. *The Zoist*, I, no.4 (1843), pp.364–5.
5. Bray, *Philosophy of Necessity* (2nd ed.: London, 1863), p.vii.
6. [anon.] *Phrenology*, p.7.
7. W. B. Hodgson, in *Phrenological Journal*, XIX, no.87 (1846), p.127.
8. *Diary of the late John Epps*, p.170; Macnab, *Analysis and Analogy*, p.51.
9. Combe, *Essays on Phrenology*, p.335.
10. G. R. Porter, *Progress of the Nation* (London, 1843), pp.260–64; quoted in *The Zoist*, II, no.5 (1844), p.11.
11. *Hansard* [3rd ser.] CIX (1850), c.34.
12. Lord Dunfermline to Combe, 30 June 1848. NLS 7292/85–88; Samuel Lucas to Combe, 11 May 1848. NLS 7294/70.

13. Combe to Ellis, 10 April 1848. NLS 7391/403.
14. *Journal*, II, no.6 (1825), p.237.
15. Combe, *Elements*, pp.173–4.
16. *Report of the Speeches delivered at a Dinner*, p.5.
17. Combe, *Lectures on Popular Education*, p.20.
18. Simpson, *Necessity of Popular Education*, p.63.
19. *Journal*, VI, no.25 (1830), p.540.
20. Schmitz to Combe, 4 November 1849. NLS 7303/124.
21. Combe, *Lectures on Popular Education*, p.24.
22. Simpson could not resist quoting Locke's story anyway. *Philosophy of Education*, p.59.
23. *Journal*, I, no.4 (1824), p.584.
24. Simpson, *Necessity of Popular Education*, p.73; and *Philosophy of Education*, pp.52–3.
25. Quoted by Gibbon in *Life of George Combe*, I, p.59.
26. George to Andrew Combe, 20 October 1841. NLS 7379/29.
27. Cobden to Combe, 28 September 1848. B.M. 43, 660/152 and 9 March 1841, B.M. 43,660/12 (C.P. XIV).
28. Combe, *Lectures on Popular Education*, p.47.
29. W. J. Fox, quoting Mann in Parliament, *Hansard* [3rd ser.] CIX (1850), c.38.
30. Simpson, *Philosophy of Education*, p.200.
31. *ibid.*, p.206.
32. *Journal*, I, no.4 (1824), p.587; and II, no.7 (1825), p.442.
33. Simpson, *Philosophy of Education*, pp.44, 45, 110.
34. Despite the variety of infant schools, there were at least two in Bury which were identified with Wilderspin. See the *Report of a Committee of the Manchester Statistical Society on the State of Education in the Borough of Bury* (Manchester, 1836).
35. Combe to Spurzheim, 20 January 1829. NLS 7384/191.
36. For Wilderspin's role and activity in the infant school movement, see W. A. C. Stewart and W. P. McCann, *Educational Innovators 1750–1880*. (London, 1967), chapter four.
37. Simpson, *Philosophy of Education*, p.46.
38. Wilderspin, *Importance of educating the Infant Poor* (2nd ed.: London, 1824), pp.60–70; *A System of Education for the Young* (London, 1840), p.76f.
39. *Journal*, VI, no.24 (1830), p.421.
40. Simpson, *Philosophy of Education*, p.159.
41. The rules of the school formed an appendix of the *Philosophy of Education*, pp.224–6.
42. Simpson, *Philosophy of Education*, p.112.
43. *ibid.*, p.196.
44. *ibid.*, p.177.
45. Whately, reporting to Combe on Wilderspin's tour of Ireland, 16 March 1838. NLS 7248/169.
46. *Journal*, XIX, no.88 (1846), p.235.
47. *Report of the Manchester Statistical Society on the State of Education in the Borough of Salford in 1835* (London, 1836),

p.14.
48. *Report of the Select Committee on Education in England and Wales* (1835), p.16.
49. *Journal*, VII, no.28 (1831), p.116.
50. *ibid.*, p.113.
51. Simpson, *Philosophy of Education*, p.166n.
52. Williams to Combe, 20 September 1852. NLS 7330/99.
53. *ibid.*, f.102.
54. George Combe, Robert Cox and others, *Moral and Intellectual Science, applied to the Elevation of Society* (New York, 1848), p.14.
55. Williams's policy was given in a pamphlet entitled *Who should teach Christianity to the Children? The School Masters or the Clergy* (Edinburgh, 1853).
56. Combe to Alexander Ireland, 6 June 1847. NLS 7391/91. Also Andrew Combe, *On the Introduction of Religion into Common Schools* (posthumously: London, 1850), p.2.
57. The pamphlet, a reprint from the *Scotsman* for 7 April 1849, is in the Cobden Papers (XIV), B.M. 43,660/202–03.
58. Combe to Ellis, 5 April 1849. NLS 7391/701. Extracts from the four annual *Reports of the Edinburgh Secular School* (1850–53) are found in an appendix of William Jolly's *George Combe as an Educationalist*, p.680f.
59. Seedair, *A Tract for all Time. The Christian or true Constitution of Man*, p.17.
60. *Report of the Select Committee*, p.122.
61. *ibid.*, p.172. A handbill entitled 'High School of Glasgow – English Department, conducted by Mr Dorsay' was sent to Combe in October 1836. NLS 7238/133.
62. *Journal*, X, no.49 (1836), p.179.
63. Simpson, *Philosophy of Education*, p.119.
64. *ibid.*, pp.124–5.
65. Clark to Combe, 20 January 1851. NLS 7313/68.
66. Stewart and McCann, *Educational Innovators*, p.338.
67. W. E[llis], 'Classical Education', in the *Westminster and Foreign Quarterly Review*, LIII (July 1850), p.409.
68. From a prospectus on education, quoted by E. K. Blyth, *Life of William Ellis* (London, 1889), p.92.
69. J. M. Goldstrom, 'Richard Whately and Political Economy in School Books, 1833–1880', in *Irish Historical Studies*, XV, no.58 (1966), p.146.
70. Ellis, *Thoughts on the Future of the Human Race* (London, 1866), p.133.
71. Kennedy to Combe, 11 March 1847. NLS 7286/36.
72. *Chambers Journal*, XVII, no.424 (14 February 1852), p.103.
73. *Baker's Biographical Dictionary of Musicians*, revis. N. Slonimsky (5th ed.: New York, 1958), p.1011.
74. Mainzer to Combe, 22 May 1848. NLS 7295/141.
75. *ibid.*

The Philosophy of Education

76. In their excellent study of educational innovators for the period 1750–1880, Stewart and McCann indicate the minor role of music and mention only the work of John Hullah at Exeter Hall.
77. Simpson, *The Normal School as it ought to be* ... (Edinburgh, 1850), p.7 and *Necessity of Popular Education*, p.83; also George Combe, preface to Caldwell's *Thoughts on Physical Education*, p.vii.
78. Smiles to Combe, 20 December 1837. NLS 7244/10–11. The *Phrenological Journal* (XI, no.56, p.318f) favourably reviewed Smiles's book.
79. Andrew Combe, *Management of Infancy*, chapter 5.
80. *Journal*, VII, no.29 (1831), p.206.
81. Combe, *Lectures on Popular Education*, p.84f.
82. Printed note of examination schedule in Combe Papers, dated 25 July 1854. NLS 7345/64.
83. Mott to Combe, 20 April 1847. NLS 7287/28.
84. Combe to Cobden, 7 March 1841. B.M. 43,660/9 (C.P. XIV).
85. *Journal*, XIV, no.68 (1841), pp.231–7.
86. *ibid.*, VI, no.23 (1830), p.294.
87. *ibid.*, I, no.4 (1824), pp.587–8.
88. Williams to Combe, 31 October 1848. NLS 7298/157.
89. Simpson, *Necessity of Popular Education*, p.134.
90. Combe, *Lectures on Popular Education*, p.114.
91. Thomas Huxley, *Science and Education* (New York, 1902), p.121n.
92. V. Knox, *Remarks on the Tendency of certain Clauses in a Bill now in Parliament to degrade Grammar Schools* (London, 1821), pp.65–6.
93. *Westminster Review*, n.s. I (April 1852), p.419.

Chapter IX   **The Politics of Education**

In 1836 a public debate was arranged in Edinburgh between James Simpson and John Colquhoun, the member for Kilmarnock. The audience expected a heated discussion and they were not disappointed, for the two adversaries each represented an unbending position on the issue of education. Colquhoun, a member of the Glasgow Protestant Association (which published several of his works), repeated the arguments of his most recent pamphlet, *The Uses of the Established Church to the Protestantism and Civilization of Ireland*. Simpson, whose views were perhaps even better known thanks to his appearance before the Select Committee on Education in the previous year, fell back upon the obstinate logic which he had used in the *Necessity of Popular Education*. In many ways it was the classic confrontation between sectarian and secularist, between the traditionalist and the reformer: a drama destined for many repeat performances in the years to come. Simpson probably thought that he had gained the last word by publishing an account of the debate under the title *Anti-National Education; or, the Spirit of Sectarianism morally tested* (1837); but if the illusion existed, it did not last long. The debate with Colquhoun proved that religious apprehensions over national and secular education, far from subsiding, were still rising, and similar debates held in the late forties would make this one look amiable by comparison. The phrenologists, however, were not so glum. Clinging to the belief in social progress and cerebral evolution, they interpreted things differently, for they saw many encouraging signs that a national system of free, compulsory and secular education might emerge in their lifetime.

The first encouraging development, noted by Simpson in the *Necessity of Popular Education*, was the growing publicity which educational reform was receiving from the press. The *Westminster Review*, the Edinburgh *Review* and a host of daily papers across the country all regularly devoted pages to the issue. Phrenologists noted with satisfaction that these reports frequently compared British education with that in Germany and America and found it lacking.[1] It was now a matter of national pride to set the school house in order; as Thomas Wyse argued in 1836, everyone was going in the right direction except Great Britain, 'the only country of the civilized world which does not have a national system of education'.[2] Given the variety of schools in the country, the widespread opposition to a compulsory and national system was probably inevitable. And yet the reluctance to adopt such a system was, in Simpson's opinion, no less foolish than the opposition to the formation of the London police force. In both cases there were cries against despotism and the loss of personal liberties — objections which Simpson thought entirely irrational. For his part he would not hesitate to 'send the children that infest the streets of London and other great cities to school by compulsion',[3] and an increasing number of papers and journals were coming round to the same view. A uniform and national system of education would serve to break down the 'upper class monopoly of knowledge' and the 'mock education' of the dames' schools. Describing such a system, Dr Hodgson declared that as education was more evenly diffused 'its character will be improved; it will become, at once, more practical and more liberal, more comprehensive and more minute ... the barriers which now divide ranks will be broken down by a common consciousness of dependence'.[4] Journalists everywhere were expressing themselves in much the same way: somehow Adam Smith's principle of free market competition did not properly apply to education. The issue was one which could only be solved on a national basis, for (as so many respectable papers warned) the rise and fall of the country might well depend on the quality of instruction provided for the common people. Education was thus a social responsibility in the broadest sense, and if, as the *Phrenological Journal* pointed out, society had already accepted the notion that the government must restrain a person from; 'filthy habits which induce disease and infect his neighbours', then it was equally desirable for the government to rescue people from 'gross ignorance which [leads to a disregard] of the public welfare'.[5]

Another promising development was the national recognition that at some time children must in some way be given a wider range of studies. Surely this was the common denominator of all the recent experiments in education. 'If a national system of education is to stop at reading, writing and ciphering', Simpson asserted, 'it would save much trouble and disappointment not to attempt it at all.' Lately even clergymen had begun to expand the curricula of their schools, and in 1832 the Society for the Promotion of Christian Knowledge established a committee, chaired by the Dean of Chichester, to publish text books and pamphlets on topics of political economy and the sciences.[6] Phrenologists happily noted that an increasing number of clergymen were disposed to practical education, and when he presented his education bill in 1850 W. J. Fox observed (with tongue in cheek) that the clergy themselves admitted to 'a deficiency of books' and were urging schools in their districts to invest in something other than hymnals and catechisms.[7]

The most encouraging signal of reformed education was the recent founding of experimental schools all over the country. The *Quarterly Journal of Education* thought this phenomenon was caused by the desire for 'solid and useful instruction' and by a growing contempt for the inadequate parish schools;[8] to phrenologists it was a portent of the withering away of the old curriculum and its replacement by a more practical one. The appearance of the mechanics' institutes (so frequently used to remedy the defects of primary education) in the twenties and thirties was only one aspect of this growth; there were also the technical-arts schools such as the one in Edinburgh (after 1821), and the system of infant schools started by the disciples of Owen and Wilderspin. At the same time the old church schools were finding it harder to match the development of new rivals, and in some embarrassment the government was obliged to grant more funds for the building of new parish and National Society schools.[9] In the decade beginning in 1839 the government was to contribute almost half a million pounds for 3000 new schools, even though most of this support went to schools of the established church.[10]

Government money meant government control or at least government supervision, which was another good sign for the phrenologists. They welcomed not only the increased expenditure in Britain but also the developments in Ireland as pointing the way to a truly national system of education. In Ireland, the number of students in the 'National Schools' (those which accepted both government funds and inspectors on a regular basis) had risen from over 100,000 in 1833, at the time of the first official report, to over 400,000 in 1845.[11] Ireland was looked upon as a preview of the good things to come, for it seemed that in the National Schools even the religious problem was well on the way to solution. Some scriptural readings were permitted as long as they were not controversial; one afternoon was set aside each week (not on Sunday) for the doctrinal education of individual students; the Irish Commissioners included Archbishop Whately and Dr Murray, the Roman Catholic Archbishop of Dublin, who appeared to get along quite well together.[12]

Lastly, the prospect for national secular education was brighter simply because of the general momentum of reform. Not long after the political adjustments of 1832, politicians (led by the Chancellor, Lord

Brougham) began talking seriously about major adjustments in education as well. Phrenologists had a high regard for the 'political wisdom' of Lord Brougham; he recognised the usefulness of the infant schools and he went so far as to advocate their establishment throughout the nation.[13] Admittedly there was the risk that Brougham or any one of his friends might turn the question of educational reform into a political football; but even that would be a welcome event if one or the other party eventually found itself committed, a prisoner of its promises, to a course of action acceptable to radicals. Politically the Whigs were the more obliging. When they returned to power in 1846 they came determined to raise teacher qualifications, to replace monitorial teaching, and to put as many teachers as possible in the pay of the state.

Phrenologists viewed all these developments in the thirties and forties and concluded that the next installment of reform, while it might not accomplish everything, was none the less certain to bring needed changes. Naturally they were anxious to hasten the process, and as they gained reputations as reformers they began to feel that they had a real part to play in the making of a national system. By the late 1840s they no longer considered themselves mere publicists of secular education — most of them had done at least this much ever since they became phrenologists. It was not a question of what education must become: they believed they had already found the answer to that problem. It was, rather, a matter of prodding an able politician or national figure to assume the leadership of a campaign — a crusade to catch the imagination of the country and translate educational theories and experiments into the law of the land.

In their search for such a champion the phrenologists turned (and returned) to three possibilities. There was, first of all, the possibility of a measure in parliament, resolutely sponsored and designed to attract liberal support; there was the alternative, decidedly less attractive, of extraparliamentary agitation; and finally there was the good example which might be given at the royal palace. None of these alternatives was to work, either separately or in conjunction with the others; but as they absorbed a considerable amount of the phrenologists' time and energy, each must be examined in turn.

1

As the indications of the solution to the education question multiplied, phrenologists first searched their own ranks for a possible leader. Inevitably James Simpson was a candidate: ever since his testimony before the Select Committee in 1835 he enjoyed a number of powerful connections — or 'parliamentary supporters', as Hewett Watson jealously called them. Simpson was on good terms with Brougham and Wyse and he was widely appreciated for his forceful speaking which, by the late thirties, involved more public lectures on education than on phrenology. There was never any doubt where he stood. 'You say you are "disposed" to become a radical', Simpson quoted Combe in 1839; 'I *am* a radical on the education question.'[14] For a while Simpson was indeed the darling of Scottish radicals and his lawyer friends assumed that he would have a role in drafting separate education proposals for Scotland. But Simpson was not in parliament and he had no burning desire to get there. In spite of his background in law he seemed quite uninterested in a parliamentary career or in any other commitment which might remove him for long periods of time from Edinburgh.

His friend Thomas Wyse possessed all of Simpson's merits in abundance and had been in parliament since 1830. Phrenologists were tempted to think that in Wyse they had found a champion. A man of some ability, not fiercely partisan, he might introduce the proper bill and see it through to victory. Wyse very early won the attention of phrenologists. He was sympathetic to science and he was proud of his correspondence with the Combes. Moreover, although he had been O'Connell's deputy in the struggle for Catholic emancipation, Wyse had the notable talent of being able to transcend religious issues — a talent likely to give him great advantage in the coming debates on education. Combe was never certain whether Wyse was a Catholic or a protestant: his speeches were always 'so above sectarian considerations'.[15] It was apparent from his essay on *Education Reform* (1836) that Wyse entertained sensible ideas on the subject, and it was known that in writing the book he relied on the counsel of his friend James Simpson. However by 1840 Wyse was quickly growing tired of the fight: too much had happened to discourage him. The programme which he devised for Irish education had come to naught, partly because of its timing (Wyse tried to push it when his colleagues in the house were absorbed in the Reform Bill of 1831); in 1839 his suggestions for an education commission were circumvented by the crafty efforts of Lord John Russell. Wyse proved a fine ally and a dedicated reformer — at Westminster he was known as the 'member for education' — but after his harsh fortune in the thirties he was unwilling to place himself at the head of parliamentary agitation. Slowly he withdrew to the background, gradually losing interest in domestic matters, and finally he removed himself from the scene altogether, first as a member of the Board of Control for India and later (1849) as ambassador to Greece.

While the phrenologists would have welcomed Wyse's leadership, they did not rule out the possibility that some one else might take his place. There were a number of parliamentarians, friendly both to phrenology and to the cause of secular education. There was Joseph Hume, who corresponded with several phrenologists and who valued the views of Combe and Simpson. There was also James Abercrombie, a Whig, a phrenologist and a confidant of the Combes. Abercrombie (1776–1858) had entered the commons as early as 1807, and (with Francis Jeffrey) had represented Edinburgh in the first reformed house. He was a politician to his fingertips. Before 1832 he occupied a number of Scottish offices and afterwards became one of the touchstones of party strength. He joined Grey's government as master of the mint, a position which did not interest him greatly as he was more anxious to speed the process of political reform through popular education. He was not afraid of 'adding to the strength of the people' by completely destroying a system of education which served only to 'perpetuate the idea of the English gentleman'.[16] His actions were almost as bold as his words. Abercrombie was a promoter of the United Industrial School 'for training destitute children' in Edinburgh; he stood ready to support any institution 'which advanced the mental development of the people'. But perhaps Abercrombie was too vociferous. There were times when he was so critical of the established church, and so outspoken on the education issue, that even democrats like Hewett Watson were astonished. Unfortunately this enthusiasm came a little too late in his career and was constrained by his election as speaker of the house in 1835. He was then obliged to mince his words. In 1839 he entered the upper house as Lord Dunfermline; thereafter bad health and advanced age brought him to the sidelines. Even so, his sense of timing and his singular focus were never impaired. Through the late forties, when the idea of nondenominational education gained further ground, Dunfermline wrote to Combe or Simpson at least once every week, advising on tactics and personalities. He would have no interference with the agitation, no blurring of its objectives: whether the crusade was raised in parliament or maintained outside, it must be singleminded and unconnected to other reforms.[17] Of course the problem of leadership remained, and, as Abercombie confessed, it was too late for him to do anything about that.

This left the field open to the amateurs. Among the candidates, one of the most skillfull — and probably the most ambitious — was Duncan McLaren, a member of the Edinburgh Town Council. McLaren (1800–86) was not related to the editor of the *Scotsman*, although both were phrenologists and both were anxious to undermine the religious control of Scottish education. As treasurer of the city government in the forties, McLaren gained a base from which he could influence municipal politics for almost twenty years. It was not until 1865 that he entered national politics as a member for Edinburgh, but for a long period before that date few men in Midlothian wished to cross McLaren. He wrote to Combe frequently, discussing various reports on education or his correspondence with Horace Mann; he kept in touch with a wide variety of people. The problem, however, was that they were mostly Scots. McLaren's ambition, his connections, and his careful attentions were centred in Scotland; in England he was an unknown quantity.

The reverse was true of Richard Whately. He never intended to enter the arena of politics: he could not imagine himself in the House of Lords, and he often protested that he had no liking whatever for party politics.[18] But Whately's problem was that no one believed him. His contemporaries expected great and partisan deeds from him. He had argued boldly for the abolition of the death penalty, for the removal of Jewish disabilities, for penal reform, free trade and an international currency; there was hardly an issue on which the former Oxford don had not written or spoken. And he continued to express himself after he moved to Dublin in 1832, where it was widely assumed that he had gone to oversee a Whig scheme of education for that country. Indeed, as things turned out, Irish education was probably the most political job Whately was ever expected to do, and it was one which effectively satisfied, once and for all, any desire he might have felt for hobnobbing with politicians. Otherwise Whately's politics (as distinct from his political philosophy) were quiet and not a little amateur. To many of his admirers and opponents, however, Whately's repose seemed calculated and unreal. They imagined that at any moment he, an archbishop, would astound the nation by emerging as the chief spokesman for a national system of free and nondenominational education.

True, Whately had much experience in education: he had written text books in political economy and he actively supported experimental schools. But it was difficult to judge accurately the extent of Whately's political power, and not every one thought, as Dr Arnold did, that once Whately put on the episcopal vestments his fame stood still. It was not generally appreciated that Whately enjoyed almost no influence either in the Lords or among Whig ministers. In fact he attended only the debates on Irish and United Church questions and when in London spent much of his time pursuing the findings of phreno-mesmerism, homoeopathy and magneticism. It possibly occurred to Combe and Simpson that Whately might, on the basis of a successful education scheme in Ireland, step forward to advocate a similar plan for England, but there was nothing they could do to disengage the Archbishop from his Irish duties, his writings, and his devotion to mysterious sciences. And so it was that the phrenologists realised (sooner than others) that the Archbishop was not a political animal after all.

The search continued. As late as 1848 Samuel Lucas, the Manchester journalist, politician and co-founder of the Anti-Corn Law League, wrote to Combe that he was 'constantly on the lookout' for someone to carry educational reform to victory at Westminster. In view of the events of the last few years, it was not surprising that Lucas expressed the hope that Richard Cobden might still be persuaded to 'place himself at the head of the movement'.[19] Actually the phrenologists had long counted on Cobden. They had noted his talents (and his noble head) long before he gained a national reputation. 'I wish you could see Cobden before he goes to Italy', chattered Sir George Mackenzie in 1836. 'He is just your man.'[20] Cobden was indeed the ideal champion. He spoke of the 'revolution in the schoolhouse' which must precede that of the franchise, and he obviously appreciated the value of technical and commercial education. George Combe lost no time in getting acquainted with the former calico-maker, and he began at once that warm correspondence which for twenty years marked a durable friendship. By the mid-1840s they were exchanging letters weekly, and Combe apologised for the frequency with which he 'added to the affliction' of Cobden's massive correspondence.[21] Combe wrote because he regarded letter writing as a duty; moreover Cobden was a particularly important friend who shared not only the same opinions on education but also a devotion to the new mental philosophy.

Cobden was converted to phrenology before he began corresponding with Combe in 1836. He first came to the philosophy by reading the *Constitution of Man* — in his case it was not the incidental result of medical studies at Edinburgh. Morley notes that on Cobden's return from America in 1833 he 'amused himself' by analysing the character of his fellow passengers, by fitting them into different physiological and phrenological types. And while in America Cobden was annoyed (as Combe was later) by the large faculty of *self-esteem*, which he thought the 'national feature' of the American people.²² As a young man Cobden wrote a comedy called the *The Phrenologist*, in which he poked fun at the more popular aspects of head-reading. And yet Cobden took his phrenology seriously. In 1835 he was instrumental in forming the phrenological society in Manchester and when he first wrote to Combe it was to ask him to visit the new society and deliver a series of lectures.²³ For years he liked to recall (with an apparently unaffected reverence) his first meeting with the author of the *Constitution of Man*, who examined Cobden's head and declared that if the free trader had been born in the middle ages he would probably have been a monk, for his *veneration* was large. Cobden regarded the analysis as 'a real triumph for phrenology' because Combe could never have formed his impression 'from anything you had seen or heard of me – and I do have a strong religious feeling'.²⁴ Mental philosophy was an integral part of the Combe-Cobden correspondence, and both men believed that phrenology's acceptance was only a matter of time. The problem of any new science was that people resented having their 'preconceived notions superceded'.²⁵

Combe was faced with quite a different problem in 1846. He wanted to secure from Cobden the definite commitment that the free trader would soon take up the banner of educational reform. It was only reasonable to suppose that Cobden would do so. When he began his career in the moor country of Lancashire, his speeches concerned the education of the young, and some of his earliest letters involved arrangements for twenty children from an infant school in Manchester to visit Sabden – as an example and inspiration for local teachers. At the end of the Corn Law agitation, Cobden was permitted a long vacation in order to rest his faculties, but wherever he went Combe's letters followed. Combe's hints were not subtle. The time was now ripe: Combe saw 'a vista opening which affords the brightest prospect that has ever greeted my vision since I became a reformer', and Cobden had a place in that vista.²⁶ The visions of future triumphs were always mingled with congratulations for past achievements. Cobden was to be flattered into a new assignment. At one point his qualities of leadership were compared to those of Napoleon (while Bright was only ranked with Marat); in another letter Combe praised Cobden for 'interpreting Nature correctly and inducing men to obey her precepts'. Cobden was the new Saint John, calling in the wilderness of the old political economy, and the multitudes heard his voice.²⁷ Combe trusted that he would never see Cobden holding high office, but that did not mean that there was no work left for him to do. Actually Combe's flattery was a bit hypocritical, for he doubted that the repeal of the Corn Laws was in the best interest of the country, and privately he predicted that 'even unlimited free trade will only deepen the misery of the working classes, and render a crisis more certain' by creating a glut in the market.[28] Even so, his high praise seemed to have the desired effect, for in July 1846 (shortly after the repeal of the Corn Laws), Cobden wrote that 'education is the only public matter upon which I should now be disposed to put on my armour for another 'seven years'' war'.[29]

In spite of his assurances to Combe, Dunfermline and others, Cobden was in no hurry to take charge. He returned from his travels refreshed but showed no signs of donning his armour. Morley has remarked that Cobden, while a man of action, never pledged himself to a cause which did not have a good chance of succeeding, either in legislation or in the public imagination.[30] And so it was with education. In the late forties, when it might have seemed that the country was ready for a measure on secular education, Cobden hesitated. Unlike other phrenologists he did not think that a motion on behalf of secular education, even if it were rejected at Westminster, would stir the country.[31] Whatever promises he may have made to Combe, Cobden had neither the fervour nor the time for planning public agitation. Cobden was (as he admitted himself) 'cautious to a fault'. In 1848 he was extremely reluctant to raise an issue on which his liberal supporters were already divided; 'To raise National Education at this moment', he said, 'would be to throw a bombshell into their ranks'.[32] At the same time he held out the contradictory promise of 'connecting the education question' to any new reform movement in which he took part.[33]

In fact Cobden did not know what direction the 'new reform movement' might take. After 1846 he drifted from one idea to another and he had no real objective in mind when he asked his old free-trade associates to join him in a new round of reform.[34] For three years Combe and Simpson tried not to lose patience with Cobden. They talked about the investments they had lost in Illinois railroads or the 'war tendency' of British aristocrats or Combe's scheme to give a 'national parliament' to each of the three kingdoms. Combe praised Cobden's temperance (he took only one glass of wine daily) and Cobden acclaimed Combe's honesty. Finally, however, the compliments wore thin. It became clear that Cobden had ruled out the possibility of an early move on education in parliament. It was not only that he feared disrupting his fellow liberals any further; he was increasingly coming to the belief (which Watson had long ago reached) that another extension of the franchise was equally if not more important.[35] With the tumult of Chartist demonstrations and European revolution still ringing in his ears, Cobden concluded that people were more likely to demand the vote before they were properly enlightened. Besides, the liberals were already in such disarray that it was better to wait until the divisions healed. In the meantime there were other national problems to tend to, and Cobden, who reminded Combe that he could 'only do one thing at a time', decided tentatively that the question of national finances would next absorb his energies.[36]

These were all mere excuses when placed alongside Cobden's principal reason for holding back. He was sensitive to the requests of his friends, and he was willing to take up education sooner or later; but at the moment the very word 'secular', particularly as a modifying word before 'education', frightened him. Cobden did not welcome the prospect of being swept into the whirlpool of religious argument — which was the inevitable result if he accepted Combe's premise that one of the aims of national education must be to persuade the clergy to 'modify their old creeds'.[37] To Cobden this sounded like a futile attack on doctrinal Christianity, and he was loathe to see such tactics become part of the education campaign. Reformers, he thought, should stay clear of the word 'secular'; for Combe, Ellis and Simpson to demand the removal of the Bible from the classroom made as much sense to Cobden as 'pressing for the *introduction* of the Koran to every school'.[38] Did it really matter how they identified their goals in education, as long as they all meant the same thing — 'to teach the people something necessary for their well-being, which the ministers of Religion do not teach them'?[39]

To the Combeists it mattered a great deal. Whenever they spoke of education, 'secular' was the most fitting adjective. Ellis insisted on it and so did Mainzer, who thought it was 'only monastic persons and bigots' who took the word to mean 'godless, worldly or irreligious'.[40] They continued to exert whatever pressure they could on Cobden, and they continuously begged Combe (whom they knew to have great influence on Cobden) to win him over by personal appeals. Through the late forties and into the fifties, Cobden resisted their pressure, explaining his 'tenderness towards religious sensitivities' by reminding his friends of his large endowment of *veneration*. Some, like Lord Dunfermline, reacted bitterly to Cobden's intransigence by arguing that perhaps Cobden would not have been the best leader after all; but Combe never gave up. Writing about education in the *Westminster Review* in 1852, he called on Cobden to put the keystone of all social reform securely in place. 'Having been one of the grand instruments for insuring a constant supply of food for the people', wrote Combe pointedly, 'the still more important work remains for him, to bring his great moral and intellectual power to bear on the supply of useful knowledge and moral training for their minds. Success in this will form the crowning merit of his life.'[41]

As the year 1850 began, the phrenologists were still without a parliamentary champion for national education. The lament which Samuel Lucas addressed to Combe the previous year might well have been written and sent again. No one was willing to introduce a bill. Lucas mentioned a number of men (significantly omitting Cobden) and explained why none of them dared to step forth. Ewart was interested but not very determined; Brougham was as enthusiastic as ever and for that very reason was overwhelming; Bowring was already in China. The only course open to phrenologists was to bestow their compliments upon any one who made the smallest move in the right direction. And there is every indication that people at Westminster were aware of the help phrenologists gave to radical proposals in education. In the spring of 1850, for instance, William Johnson Fox, one of the organisers of the Anti-Corn Law League and the member for Oldham, brought a bill before the Commons: the proposals were similar to those of the Lancashire Public School Association and envisaged (among other things) a system of free schools for children from seven to thirteen, provision for religious instruction outside the state-supported schools, and encouragement and publicity given to able teachers as an example to others. Phrenologists welcomed the measure even though, as Cobden predicted, there was absolutely no chance of its passing; opponents in parliament (like Lord John Russell) would have nothing to do with a measure which promoted a variety of teaching 'exceedingly inferior to any which admits of the teaching of the Bible'.[42] The Duke of Argyle (widely believed to know what he was talking about) was very distressed: he imagined that there was a secularist conspiracy afoot. Who exactly, he asked, was behind Mr Fox? The Duke knew and he wanted everyone else to know, too: he referred to a group in Scotland 'composed of somewhat heterogeneous materials' who were out to wreck the existing system of church-affiliated schools.[44]

The Duke was thinking of the phrenologists. He had never liked the moral philosophy of Spurzheim and Combe, and he liked Mr Williams' Secular School even less. He could not prove that Fox considered himself a 'happy cooperator' with George Combe, or that the member for Oldham regarded the Williams Secular School as 'a great step in education' but he guessed as much.[44] It was well appreciated in the house that certain members did not approve of changes in the curriculum of schools merely for the novelty of school subjects such as physiology or astronomy; nor did they favour secular education simply because they were bored with the claims and counterclaims of different religious groups. No: the reformers who dabbled in phrenology or spent decades corresponding with phrenologists did so because they wanted a reasonable and possibly a scientific foundation for the changes they knew to be necessary. Several of the parliamentary radicals who voted for Fox's motion in 1850 were financial supporters of the Williams Secular School, which they thought a suitable model for the whole country.[45] Their interest in such experiments was philosophical and they were ready to believe that the timing and presentation of school subjects was largely a matter of psychology. Speaking on the education issue in 1853, Lord John Russell openly acknowledged one of the leading factors in the secularist movement. He knew that many of the honourable members were inspired by George Combe, 'a gentleman no doubt whose writings ... and whose opinions are of considerable weight'.[46] Russell measured this weight in terms of a radically new curricula and natural theology, and he clearly expressed his opposition to both. But his frank words dissuaded no one. Several of his colleagues in the house continued to add the psychology of George Combe to their views on education. For Cobden the 'only definition of education' was Combe's; for Lucas the whole cause of national education owed 'more to Combe than to any other single person'.[47]

Still, it was now clear that the politicians could do nothing. Westminster was not yet ready for secular education. The measure which Fox put so decisively before the country in 1850 had failed to create much commotion — even though it may well have been the most thoughtful alternative at the moment. It confirmed the suspicion (held equally by Watson and Combe) that the commons was still 'too aristocratic, too clubbish, and bent on traditions'. It also settled Cobden's mind once and for all and brought to an end most of the pressure which phrenologists had exerted on him for years. The double failure of Fox's bill and the attempt to persuade Cobden made the phrenologists turn away from the politicians; they began to think that popular agitation of one sort or another was more likely to bring results.

2

With few exceptions, phrenologists had never been inclined to extraparliamentary agitation. Like Combe, Mackenzie, Simpson and Ellis, they were members of an increasingly contented middle class. They had always had the vote, and they enjoyed a vague belief that, given time and effective leadership, almost anything could be got from parliament. The prospect of having to resort to, or depend upon, the now established techniques of popular agitation was rather upsetting. 'Where are the *educated* men we need to assist us?' asked William Ellis in 1850. 'Are we now obliged to enlist unlettered fishermen?'[48] The lament was all too characteristic of a man whose unbridled faith in classical economics made him condemn strikes and combinations as the 'folly of the artisan class'. And yet his fears, so typical of his class, were common to other phrenologists. Seeing themselves as educators, philosophers or scientists, they did not particularly feel that they should enter the arena of politics themselves; to some extent they all shared Combe's position as a 'teacher of teachers' whose views on education influenced intelligent minds.[49] Their trust in the great mass of population was always tentative, for the common people were too often led astray. 'In America', Combe once told Samuel Lucas, 'I saw this same class of men, while ignorant, was the dupe of the demagogue or the priest.'[50]

Not all phrenologists agreed with Combe on this point. They were willing to trust his dupes and Ellis's unlettered fishermen. In any event, the population (franchised or not) was waiting to be used as a great lever with which to pry a satisfactory response from parliament. Simpson was quick to recognise the possibilities. In the *Necessity of Popular Education* he spoke of a 'war on ignorance' to be waged by all men; he foresaw a campaign in which public opinion would be essential.[51] The same notion appealed to others. Charles Bray (who by 1848 was the proprietor of the Coventry *Herald*) and Robert Chambers (of Edinburgh publishing fame) were eager to marshall public support, regardless of class. And of course there was always Hewett Watson, who Simpson thought was born a whig and christened a democrat: Watson, too, was anxious to see agitation on the broadest possible scale. He proposed setting up a 'national Reform Association' with branches in every county, its goals democratically determined by the results of 'censuses of national opinion'.[52] While there might be room in the Association 'central committee' for a few high-minded reformers (Watson meant to include Cobden), its membership was to cut across the parties and, if possible, to avoid politicians. In effect what Watson proposed was a super version of the old Anti-Corn Law League: updated, better organised, geared to public opinion and to the widest possible membership, a permanent force in British society. To Ellis the whole idea was a waste of time, even though Watson offered to 'steer the organization in the direction of national education'. Combe was a bit more indulgent, but suggested several characteristic alterations. If the Association was really to work, Combe said, it must include a few lords and knights; it would also help (in this Chartist year of 1848) if the Association stated that 'honorary titles and private property are ... sacred, and that justice to all does not render it necessary to [unmake] them ... ".[53] It was not the sort of amendment which pleased Watson. At any rate, Combe was obviously unwilling to lend his support or commit his friends to a reform organisation which, however highly motivated, was not thoroughly respectable.

While Watson's plan never matured, an alternative arose in Lancashire. In July of 1847, following the decisions made by the Privy Council committee on education during the previous year, a group led by Samuel Lucas issued a *Plan for the Establishment of a General System of Secular Education in Lancashire*. At this moment phrenologists were still hopeful that Cobden would bring secular education before the commons, and they therefore regarded the activity in Lancashire as a sideshow to the real struggle expected at Westminster. Samuel Lucas (1811-65) saw things in a different light. He hoped to carry the education issue into parliament in much the same way as the Corn Law had been carried – and conquered. Lucas hoped for a system such as the one promoted by Horace Mann in Massachusetts; as a phrenologist he hoped the 'practical curricula' would 'challenge the faculties at the right time'. He sent Combe a draft of the *Plan* before it was actually published in Manchester, and he discussed with Combe a definition of secular education.[54] Lucas served on a committee of seven, five of whom were Scots and friends of Combe; the committee met several times with Combe and Dunfermline before their plans were fully revealed. And as the organisation gained strength, it occurred to Combe that Lucas was after all a remarkably sensible man. He expected – and wanted – people of all ranks to join the campaign, but he insisted on a clear-headed and respectable leadership. Provisions were made for 'organising local committees of workingmen' who might be 'called out in force at a moment's notice' and whose contribution Lucas fully expected to control as so many reserve divisions.[55] This was the most effective (and the safest) way of proceeding, for although Lucas wanted to believe in the 'good intentions of the people', he could not accept Watson's assumption that 'public opinion is always sound'. By 1849 there was no turning back; the Lancashire Association had equated 'national' with 'secular' education, and they were planning to rouse the people, county by county, until finally Westminster took notice.

A number of Manchester phrenologists were at the helm of the Lancashire Association. In addition to Samuel Lucas and Cobden (who was abroad when the Association was formed and who returned to play a distinctly minor role), there was also Robert W. Smiles, one of the Secretaries of the Association, and Milner Gibson, a member for Manchester and president of the Association for the Repeal of Taxes on Useful Knowledge. These men argued that parish-run schools were inhospitable to modern education and they demanded new schools, supported by local taxes and managed by locally elected bodies. No doctrinal teaching was to occur in the new schools, which were to emphasise 'practical subjects' such as composition and the natural sciences.[56] If some parents insisted on 'sectarian teaching', their children were to receive such instruction at home or in the parishes. None of these suggestions was original to the Lancaster Association (which looked to Massachusetts for a model) or to the phrenologist members of the Association. Yet it is interesting to note that when the Association brought its case to parliament in 1850, it did so by sending each M.P. a copy of Andrew Combe's *Religion in the Common Schools* — or, more accurately, a pamphlet of extracts from Andrew's writings on the subject, edited by his brother George.[57] In fact, Alexander Russell, editor of the *Scotsman*, confided to Combe in 1849 that some people refused to take part in the Association because they thought it was devised to 'introduce phrenology as a subject' in the reformed national schools.[58] The Association never considered the idea and did not bother to deny the rumour. The suspicions were again raised in 1851, however, when the Association established a secular school in Manchester patterned after the Williams Secular School in Edinburgh.

By the end of 1850 almost everyone was obliged to take the Manchester-based campaign more seriously. The Lancashire men, with their expereince in free-trade agitation behind them, had apparently hit upon an effective way of raising support for a comprehensive plan of secular education, first in their own county and lately throughout the nation. Their activity encouraged reformers everywhere and undoubtably prompted Fox to make his move. The fate of his motion notwithstanding, the optimism of the Association remained; late in 1850 the name of the Lancashire Association was confidently changed to the 'National Public School Association'. Combe was satisfied that the great cause was at last off to a good start, and he shared Lucas's hope that the Russell ministry could be stampeded into an acceptable measure. In his capacity as a 'teacher of teachers', Combe offered a large number of his pamphlets on *National Education* and *What should Secular Education embrace?* to Dr Hodgson and the Public School Association; he had purposely stored these pamphlets for distribution at the right moment, when a national campaign required guidance.[59] The movement continued to gain momentum in 1851. Papers in northern England and in Scotland reported large working-class meetings which issued demands for a programme of secular education in the parish schools; petitions were sponsored and sent off to London in great bulk; Combe cheerfully wrote that 'even in Orthodox Scotland there is a *living interest* among the people'.[60] The agitation was all the more gratifying in that it showed no sign of getting out of control. The timing, too, had proved right; in the words of one correspondent, it was important to raise the issue before 'Albert's china shop in the park' (the Great Exhibition) either distracted people from social reform or lulled them into complacency.[61] The obvious challenge for the Association was to maintain public enthusiasm at a high pitch for a fairly long time — Lucas, thinking of the recent war against the Corn Laws, estimated a period of up to three years — and to do this they would have to resolve the problem of money. It was here that the 'teacher of teachers' came to the rescue.

Late in 1851 Lucas reckoned that to continue the campaign, a national organisation, with headquarters in the capital, was absolutely necessary to prove the viability of the cause. 'If you want to set a fire in this country', Lucas wrote to Combe, 'you must kindle it in London; and if you make a fire in London, it must be a hot one.'[62] Lucas put the cost of the blaze at approximately £1,500 per annum, which met the expenses of several secretaries, lawyers and printing, with virtually nothing left to spare. For almost one year of the Association's existence, these expenses were largely borne by a man who preferred Combe as the intermediary. That man was Edward Lombe. Combe began corresponding with Lombe, or rather with Lombe's secretary W. E. Hickson, in the spring of 1850. At that time Lombe realised that the 'people's college' of Norwich, an institute of technical training which he established for the sons of artisans, could no longer continue. Lombe had lost over £1,000 in the venture, but its failure neither

The Politics of Education 213

ruined him financially nor did it break his enthusiasm for educational reform. Edward Lombe (d. 1852) was the descendent of Sir Thomas Lombe, who in the early 1700s had smuggled silk-throwing machinery into the country from Italy, thereby freeing Britain from the Italian silk industry and making a huge fortune for himself. Lombe still had at his disposal a large Norfolk estate of some 16,000 acres, divided into sixty-eight farms whose annual rent brought over £15,000.[63] Whatever his loss in the 'people's college', Lombe could well afford to continue his experiments.

Combe suggested making a gift to the Williams' Secular School, but Hickson's master was no longer thinking in terms of funding progressive schools. He lost confidence in experiments of this sort; if his college at Norwich had failed, what guarantee of success was there for a school in Scotland? Instead Lombe was attracted to the idea of popular agitation on a national scale, and he decided that the Lancashire Association was a useful tool. Moreover he was convinced that the author of the *Constitution of Man* and *Moral Philosophy* exercised great influence in the councils of the Association, and he determined to make Combe his agent. Combe was not unwilling to accept the role of middleman between a gruff and generous patron (who now lived in Italy) and the growing Public School Association. Combe was undoubtably motivated by the hope that he too might share in Lombe's *largesse*, and indeed modest sums were eventually designated to Combe's allocation alone. Of the first amount placed at his disposal, Combe assigned half to support working-class meetings in Edinburgh and the other half to the Williams' School.[64] Beginning in the autumn of 1850 and continuing well into the next year, Combe was the 'comptroller' of several transactions between Lombe and the Association; after making the necessary arrangements with Lucas and Smiles, Combe transferred the first gift (£500) in time to pay for the great public meeting which the Association held in Manchester in January 1851. Smiles and Lucas were delighted: the Association would easily survive as long as Lombe's generosity continued. Indeed, it must have seemed that Lombe was willing to commit his fortune piecemeal to the project; he even contemplated buying seats in the commons for Hickson and Combe, in order that they might introduce precisely the bill they desired, without having to rely on someone else, when the time came.[65]

If Lucas and Smiles had ever imagined that Lombe's money was truly a free gift, they were soon enlightened. Although Lombe addressed the Association only infrequently himself, he was not slow in directing Combe to make known the conditions which accompanied his gifts. As far as tactics were concerned there was no problem. They all agreed to emphasise education as a 'popular cause' which they hoped to encourage by arranging a 'monster petition' signed by 100,000 artisans in the Manchester-Salford area — although Smiles doubted that the two boroughs contained that number of adult males.[66] Disagreement arose, however, not over the number of signatures but over Lombe's hostility to the Association's clerical opponents. Lombe understood the whole education issue in terms of class and clergymen: on the one side was the mass of the population, and on the other were the aristocrats and the church, together with 'those great pests, the dissenters', whom he blamed for the collapse of his Norwich college.[67] For Lombe, compromise with the defenders of church-affiliated schools was entirely out of the question. While the Public School Association saw the 'monster petition' as a forceful expression of popular will, Lombe saw it (and offered to pay for it) as a means of 'smashing the fradulent church schemes'. Lucas, Cobden and Smiles did not share this sentiment, exasperated though they certainly were by the activity of a rival Manchester group. They wanted secular education, to be sure; but they could not afford to bait the religious party much further.

An explosion was inevitable. Lombe did not like the reports, which came to him from Combe, of the Public School Association's feeble attempts to conciliate the church party in Lancashire. Lombe's impatience was enormous, and so was his self-delusion. He regarded himself as the remaker of his country and as the greatest reformer since Bentham. The new reform machine in Manchester was his: he had purchased it, and he now intended to turn it into a political derrick with which to destroy the hereditary peerage, the established church, the power of the courts, and the private schools. Lombe was accustomed to thinking this way; after all, he had always found that every man has his price, and once the Lancashire men were given more money they would take a suitably hard line with any one who defended church schools. This they were not ready to do, or at least they showed no signs of becoming as vehement as Lombe wanted them to. Finally Lombe (who warned that men 'must take my opinions with my dollars') called on Combe for an explanation. Combe was in no position to account for the deeds of Smiles and Lucas, but he took the opportunity to send a message whose meaning was unmistakable. In clear language Combe declared that he was long annoyed by Lombe's attitude: Combe was not a mere servant bound to obey great orders; he was an ally, a partner, an 'independent power', and it was quite enough for him to see that Lombe's gifts were not lost or wasted.[68]

Lombe was not prepared for this declaration of independence. He very soon decided that if the ties which connected him to Combe and the Association were to be severed, he would do it himself. Lucas and Smiles regretted the end of so profitable a connection but they were not surprised, for they thought Lombe an 'erratic gentleman'. By the end of 1851 the connection was entirely broken and the gifts ceased. While the phrenologists never had the chance of meeting Lombe personally, they were now certain that his *self-esteem* and *combativeness* were unduly large and likely to reap him nothing but disappointment. He imagined himself a great reformer, whereas he was only a hapless manipulator, wildly hoping to provoke cataclysm in Britain from the opulent safety of his villa near Florence.

The National Public School Association did not long survive, but its demise was not due to the end of Lombe's money. The real stumbling block, of course, was the continuing religious furore. If national schools were established, was it possible to have textbooks to every one's liking? If Bible reading remained part of the instruction, who was to choose the extracts? How much time should be given to children of minority creeds in any parish? How were taxes to be allocated fairly to the educational needs of any district? It was almost as difficult to list the problems as it was to solve them. The Association, whose membership was by no means exclusively secular in spirit, was obliged to watch its step. The most fundamental question concerned the meaning of the word 'secular', and the focus of this issue was the practice of Bible reading in the schools. Before the Lancashire Association was established, Lucas, Smiles and Combe all agreed that it was better to discontinue Bible reading altogether; they adopted Simpson's argument that the common use of Scriptures, as an aid to reading and spelling was both impractical and irreverent. Properly understood, secular education was not godless at all: it 'rescued the Bible from degradation and perversion' and assured its unique position during 'a day of leisure and piety'.[69] The definition did not soothe Cobden's conscience. He remained frightened by the use of the word which, in the 1840s and fifties, was a synonym for atheism. Cobden may well have remembered the debate over Fox's education bill in 1850, when one member asserted that Fox himself studiously and suspiciously avoided the word.[70] The same fear was at work within the Association. Some members were afraid that if their goal in education was always described as secular, they would stand no better chance with the public than Fox had in parliament. George Jacob Holyoake later commented that the Association was neither secularist nor pretended to be so; but this absolution (from one who considered himself a professional secularist) came too late to help Lucas and his colleagues in the early 1850s. The fact that most of the leaders remained 'secularists' — that Lucas and Smiles continued to think it necessary to 'give people perfect security against sectarian teaching' — was probably enough to seal the fate of the Association.

In December 1851, the National Public School Association held a conference in Manchester. It was well attended and, thanks largely to Lucas and Milner Gibson, remained firmly and officially attached to a system of free schools 'employed to impart secular education only'. Accepting the invitation of the School Association, George Combe came to the conference and, in the name of national secular education, he delivered his last public speech. He thought that the enthusiasm of the delegates was considerable, but that was little compensation for the strong leadership he felt the Association must have to succeed. Like Carlyle, he had searched for heroes and captains and found none; he came to Manchester expecting no permanent results, and this time he was not disappointed. An uneasiness pervaded the Conference. There was the feeling that the Association had argued its case before the nation, but to no avail. The membership had ceased to grow by leaps and bounds. The religious impasse was daunting. Finances were now uncertain. Even the monster petition proved embarrassing: it gathered little more than half the projected number of signatures.[71] The deputation which the Conference sent to Downing Street in January 1852 was a pitiful episode; the interview went as expected, with government ministers, and later *The Times*, congratulating the men of Lancashire for their concern. Through 1852 a sense of hopelessness enveloped the men who planned to make a veritable revolution in the schoolhouse. To the leaders of the Association, and to their phrenologist friends, the lesson was now clear: they had signally failed to overcome the religious issues and to carry public opinion for more than a few months.

The performance of the Association was, in Combe's opinion, all a matter of leadership. If Cobden had come out more forcefully, public support might have been wider and more permanent. Who outside Lancashire knew the merits of Lucas, Smiles and Gibson? Combe never believed that these men, while philosophically sound, really measured up to the demands of a long public campaign. Ellis, too, shared Combe's suspicion; before the 'deplorable confusion' of the Manchester Conference he felt that the agitation must soon collapse for want of 'clear-sighted and earnest' leadership.[72] If the campaign was to be truly popular, its leadership could not rest with a relatively unknown and provincial cabal. It had to rest instead with a single man who was respected everywhere in the country for his personal vigour, his strength of character and for his high-minded approach to reform. It was pointless to borrow the talents of men like Mazzini or Kossuth — which is what the secretaries of the Manchester Conference proposed to do; a native son was required, and one who knew the people he must lead. As long as the National Public School Association had any chance of success, Combe thought that man might be William Lovett.

The phrenologists had known Lovett for some time. After the Chartist demonstrations in the spring of 1848, Ellis reported from London that the Chartist movement was hopelessly divided and that those of the 'moral force' persuasion might in the future prove welcome recruits to the campaign for national education. Ellis knew what he was talking about. Already he had worked with the Chartist leader William Lovett in preparing a 'suitably secular curriculum' for the National Hall School in London; besides, Lovett made his priorities known to all. In his 'Proposal for a General Association of Progress', the first of his 'social objectives' was the 'general education of the Whole Population' — the very words implied a system of national schools. Through the 1840s Lovett had the support of phrenologists at every turn. Ellis introduced him to the study of 'social science', which they then promoted at the National Hall; Dr John Elliotson, the London phreno-mesmerist, encouraged Lovett to include lessons in anatomy and physiology in the same school and proofread one of Lovett's essays on the subject.[73] Dr William Hodgson was another of Lovett's friends; he was a frequent visitor to the National Hall, where he (and Ellis) gave many lectures and philosophised with Lovett about the goals of education.

By the autumn of 1848 George Combe was also among Lovett's friends. Lovett was sent copies of the prospectus of the Williams' Secular School, which opened that year in Edinburgh; he was also asked to write a testimonial for the school for circulation in the artisan districts of Edinburgh.[74] The next year, Combe returned the favour by writing commendatory articles about National Hall in the *Scotsman*. But from a fairly early point the correspondence dealt with more than curricula and textbooks. In explaining why he was still a moral-force Chartist, Lovett revealed that his long-term objectives remained essentially political: but before the Six Points (which he had helped to draft) could become the law of the land, Chartists would have to 'prepare themselves' by acquiring the 'correct views, improved habits, and a sound education for their children'.[75] Lovett came to this conclusion fairly early in his career, and his attitude was confirmed by reading the *Constitution of Man*, which 'opened up to me the first clear ideas of my own nature, and taught me how my future would depend on my own exertions'.[76] What Lovett had found for himself in the writings of Combe, he now proposed to share with others: a practical psychology which allowed him to 'restrain passions' and stay on 'the right road' of self-improvement.[77]

Philosophically Lovett was indebted to Combe, whose friendship he greatly valued. But after Cobden, why should phrenologists have felt that Lovett was the man for the job? Usually they did not mix with public agitators, and none of the leading phrenologists appears to have been an enthusiast for the Charter. And yet they not only believed that Lovett was a man they could trust; they also dreamed of his usefulness as a leader. He was, undeniably, a national figure. He had travelled the country in the Chartist cause, and he was well-known everywhere. As a speaker he was at least as effective as any one in the Public School Association. As a moral force man he had repudiated violence as a tactic and he remained enough of a capitalist to win the full endorsement of William Ellis. Thanks to his hard work on behalf of the National Hall and its education programme for artisans, Lovett was brought within the pale of middle-class radicals, well-wishers and philanthropists, who found that he was a genial person, able and willing to cooperate with a wide assortment of respectable reformers. Indeed, as the forties gave way to the fifties, Lovett's sense of class consciousness became less acute than ever before. He was willing to submit the people to education before rewarding them with the ballot. If the National Hall — where Lovett saw to the banning of both intoxicating drink and sectarian religion — is any measure of his goal, he must have hoped to transform the working class into a reasonably educated, self-disciplined force in society. He also appreciated that this goal was psychologically more distant than he first imagined. Taken as a whole, Lovett's position was one which quite readily endeared him to the phrenologists, who throughout 1850 and 1851 deluged him with the same flattering letters and personal attention which they had once bestowed on the wary Cobden. Combe and Hodgson pressed Lovett to come out of his 'unreasonably early retirement'; Ellis, who was always in or near London, was responsible for most of the coaxing.

It was all to no avail. The phrenologists, who believed that Lord John Russell might be frightened out of his complacency by a vigorous round of agitation, discovered that Lovett was equally immovable. His response was not unlike Cobden's. He assured Combe that he wanted a transformation of national education, but in his opinion the hour had not yet come. Everyone was in such a good temper, he said: the Crystal Palace had seen to that. Besides, were he to re-enter the battlefield of politics now, his old colleagues would expect him to concentrate on the franchise question and they might not agree to help him if he insisted on education first. Lovett was biding his time. He made no promises to Lucas, Smiles or to the National Association; he declined to take an active part in the Manchester Conference of December 1851. Like Ellis he was too absorbed in conducting his school, and like Cobden he did not have the stomach for the religious squabbles associated with the education issue. Perhaps because he was closest to Lovett, Ellis was the first to realise the impossibility of gaining his total commitment to the cause. Of course Ellis had never been very fond of popular movements, and far from expecting legislation to result from public clamour, he was quite content with the clamour itself.[78] As Simpson and Combe came to realise the same thing, the optimism of the late 1840s evaporated. Popular movements were inconceivable without popular leaders; Cobden and Lovett had declined and there was no-one else in sight. And if any illusions remained, they were shattered by the Manchester Conference of 1851. The crowds were there, and so were the great men, but the noise was to have no echo. Outside the conference hall, the obstacles to national secular education remained as monstrous and as formidable as ever. Combe returned to Edinburgh feeling as though he had seen a play in which all the acting was half-hearted and everyone wore masks.

3

There remained one more path to national education, and it led through Buckingham Palace. As a possible alternative, this route was almost as old as the other two; but now that the frontal attack in parliament and the outflanking manoeuvre of public agitation had both failed, phrenologists wondered whether national education might yet begin — truly begin — with the royal assent. There was no fancy scheme involved — merely the gentle art of persuasion. If it were possible to convince the popular young queen and her consort of the merits of an eminently scientific and practical education, and if they chose a particular course of instruction for their own children, would it be long before the rest of the country followed their example? In the past members of the royal family helped to determine the artistic and cultural tastes of the nation; was it not possible for this royal family (the most respectable in living memory) to inaugurate a new style in teaching the young? Eventually, Combe thought, the new style would become irresistible, and then politicians would hasten in embarrassment to enact what they earlier dismissed as utopian and godless.

Phrenologists had no problem of entry to the court. Thanks to his appointment as physician to the Belgian king in 1836 and as one of Victoria's physicians in 1838, Dr Andrew Combe soon made the acquaintance of several members of the court, and he lost no time in converting a few of them to phrenology. As early as 1838 he reported that Baron Stockmar was taking a deep interest in the new mental philosophy and had read all of George Combe's works in German.[79] Stockmar was flattered when the Combes asked if they could help promote German educational theories in Britain, and he later assured Prince Albert that the Combe brothers were the most sincere germanophiles he knew. Stockmar was a useful connection, although not a permanent figure in the new court. Sir James Clark was. As principal physician to Victoria from 1838 until his death twenty years later, Clark was in an excellent position to see that the writings of phrenologists got into the hands of important people, and he regularly supplied the appropriate books and pamphlets to Lord Clarendon, Kay-Shuttleworth and Lord John Russell.

The most important recipient, however, was Prince Albert. The Combes regarded the Consort as a politician in his own right and a permanent one. They were convinced that once Albert saw for himself how effective a phrenologist tutor might be with the royal children, he would seek from the Queen's most reliable ministers a pledge of educational reform. The first step, then, was to prove to Albert that the best tutor he could possibly engage was a phrenologist, a man who could plan a course of studies suitable to the natural talents of children. Sir James Clark was entrusted with all the preliminaries. First Albert was given a new copy of the *Constitution of Man* (which he had already read) and Clark made a point of quoting from George Combe's latest letters on events in Germany. Combe expected an invitation to the palace as a matter of course, for he knew that Albert was not at all reluctant to mix with phrenologists. He was on good terms with Dr Elliotson, and he was also acquainted with James Deville, one of the popular 'head-reading' phrenologists in London, who loudly boasted that Albert had sent him two fine dogs and had consulted with him on the education of the royal children.[80] At any rate, Clark's efforts soon bore fruit: arrangements were made for George Combe to visit the palace in the spring of 1846.

Combe thought his presentation was successful and pleasant. After some amiable conversation with Clark, Albert and the Queen, Prince Alfred and Princess Alice were brought in and their father had each of them shake hands with Combe. Upon examining the children, Combe concluded that their moral sentiments were sufficiently strong to allow them 'to accomplish much good', and only once did his analysis of Alfred bring a dissent from the parents. Combe's hopes in connection with the visit are already clear, but lately it had been suggested that Victoria had reasons of her own for agreeing to the phrenological visit. Alfred was a backward child: contemporary rumour had it that the Queen thought him 'rather stupid' and wondered whether his intellect might be improved.[81] Combe's judgment was reassuring, and Albert was very eager to have further advice. He asked Combe to furnish him with a complete set of notes on the evaluations he had just made of the children. Combe delegated this task to Clark who unfortunately misplaced the notes, but at least Clark indicated that Albert was 'really impressed' by Combe's remarks.[82] Having made this move, it was now possible for the phrenologists to decide among themselves who was the most suitable candidate for royal tutor.

The process was long and unhurried. It must be remembered that at the time phrenologists fully expected a campaign to begin in parliament under Cobden's leadership or, that failing, a popular extraparliamentary struggle led by the Lancashire men or by William Lovett. When these two possibilities did not materialise the phrenologists saw the 'palace project' (as Clark liked to call it) in a different light. Until late in 1851, therefore, the possibility of a phrenologist tutor for the royal children was always a delightful prospect, but one of minor importance. Besides, it was already likely that Bishop Wilberforce of Oxford would hold the position, at least temporarily, and it would have been counterproductive to try to unseat him. In the meantime the phrenologists continued their influence. Within two months of George Combe's first visit to the royal family, Andrew affirmed that Baron Stockmar was won over completely: he agreed with the Combes on the sort of 'moral education' which the Prince of Wales must receive, and he began to press his views on Albert.[83] The Combes understood that a possible successor to Wilberforce was Archbishop Whately of Dublin; but they decided not to support him when the time came, because they feared that the Archbishop's 'indiscretions and vain character' might do their cause more harm than good.[84]

Unexpectedly the phrenologists had several years to find the right man, for Wilberforce stayed longer than they anticipated. The influx of phrenological literature continued. Large segments of the letters which Combe dispatched to Clark were intended for the ears of the Consort, and Clark was faithful to his duty. The propaganda found its mark and Albert asked for extra copies of Combe's *Lectures on Popular Education and Moral Philosophy* as gifts for Aberdeen and Peel; Clark secured six copies of each work for distribution by the Prince. Finally, in 1850, a new tutor was engaged. According to Clark the choice was really made by Baron Stockmar, who amidst some confusion about the position won the place for his friend, a German phrenologist named Ernest Becker. While Becker prepared to take up his duties in the autumn of 1850, Combe in turn prepared to make Becker his own pupil.

To a large extent he succeeded. Becker was presented with a set of the late Dr Andrew Combe's works, and Clark was instructed to give him lessons in physiology and anatomy.[85] During the winter holidays of 1850–51 Becker travelled to Edinburgh, where he reviewed phrenology with Combe in Melville Street. Everything possible was done to make Becker feel at home; he was surrounded by the most distinguished scientists, including Mackenzie and Samuel Brown, the atomic theorist. Dinner parties were given at which only German was spoken, and Becker, suitably impressed, declared that Edinburgh was not the New Athens after all, but the New Heidelberg. During the next three years he made several return trips to Edinburgh, and in the letters which he sent to Combe fortnightly (often reaching fifteen pages in length) he described the phrenological development of his royal charges. Moreover the new tutor proved to be an even more reliable agent than had Clark, who by mid-1850 showed signs of reluctance in his capacity as Combe's messenger in the palace.

Becker had his orders. He was to make himself useful in the education of the children and particularly of the Prince of Wales. As tutor he was to be a man of great refinement and cultured tastes, the model phrenologist whose special understanding of human aptitudes gave him a scientific pre-eminence at court. His accomplishments were, in effect, to reflect those of his science and its advocates. He was to prove that phrenologists were reasonable and practical men whose ideas on education, morality and human behaviour, far from being outrageous, were entirely sensible. If Becker succeeded in this task — if the merits of his psychological approach in education were acknowledged — the country (and particularly its leaders) would not fail to notice the results.

This was Combe's reckoning, and his consolation, after the Conference of December 1851. The frustrations of trying to extract promises from politicians and support from the public were now somewhat tempered by the calm confidence that personal contacts were likely to yield fine dividends in the future. The 'palace project' augured well for national education. Simpson and Watson thought that perhaps it was Combe's second visit to the palace in June 1850 which paved the way for Becker's appointment — a supposition which Combe himself entertained for a time. Equally important, there were indications that phrenology itself had ceased to arouse the suspicions and hostility it once encountered. Becker, a well-known and devoted phrenologist, had won approval as royal tutor; and in 1854 he was succeeded by another, William Ellis. Their phrenology was not held against them, even in those high places, and there is every reason to believe that Prince Albert appreciated deeply the phrenological advice which he sought from time to time. The policy of personal contacts may have been rather slow, but it remained more promising than any other which the phrenologists tried. In time they hoped that the lessons which Becker and Ellis delivered at the palace would be given throughout the country: the reformation of society through education would begin at the very top and work downward as 'a most welcome encouragement to the more intelligent and benevolent of Her Majesty's subjects'.[86]

Notes

1. Simpson, *Necessity of Popular Education*, pp.261–5.
2. *Report of the Speeches delivered at a Dinner*, p.19.
3. Simpson, *Philosophy of Education*, p.171.
4. W. B. Hodgson, 'Nature and Scope of the Mechanics' Institutes', *Journal*, XIV, no.68 (1841), p.236; the *Zoist* (II, no.5 [1844] p.17) later delivered the same message.
5. *Journal*, XX, no.90 (1847), p.8.
6. Minute Book of the Committee of General Literature and Education, Society for the Promotion of Christian Knowledge, I, (June–July 1833), *passim*.
7. *Hansard* [3rd ser.] CIX (1850), c.36.
8. *Quarterly Journal of Education*, I, no.2 (April 1831), pp.264–5.
9. *Recent Measures for the Promotion of Education* (4th ed.: London, 1839), p.iv.
10. Edinburgh *Review*, XCII, no.185 (July 1850), p.104.

11. *Journal*, XX, no.90 (1847), p.31.
12. George Combe described the educational promise of Ireland in *Remarks on National Education* (Edinburgh and London, 1847), pp.28–31.
13. Brougham's proposals, made in the Lords in May 1835, were summarised in the *Recent Measures for the Promotion of Education in England*, pp.4–5. His York speech of September 1835 was mentioned in Simpson's *Philosophy of Education* (p.204), and his essay 'Education of the People' was reviewed in the *Journal*, II, no.7 (1825).
14. Simpson to Combe, 15 October 1839. NLS 7252/94.
15. Combe to Simpson, 9 July 1835. NLS 7386/351.
16. Dunfermline to Combe, 3 January 1848. NLS 7292/44.
17. Dunfermline to Combe, 24 March 1848. NLS 7292/61.
18. *Reply of the Archbishop of Dublin to the Address of the Clergy* (London, 1832), p.12.
19. Lucas to Combe, 4 February 1848. NLS 7294/49.
20. Mackenzie to Combe, 6 August 1836. NLS 7240/42.
21. Combe to Cobden, 23 March 1843. B.M. 43,660/15 (C.P. XIV).
22. John Morley, *Life of Richard Cobden* (2 vols.: London, 1881), I, p.32.
23. Cobden to Combe, 23 August 1836. B.M. 43,660/2 (C.P. XIV).
24. Cobden to Combe, 1 August 1846. B.M. 43,660/58 (C.P. XIV).
25. ibid.
26. Combe to Cobden, 2 November 1847. B.M. 43,660/73 (C.P. XIV).
27. Combe to Cobden, 20 July 1846. NLS 7389/18; Combe to Cobden, 11 March 1846, B.M. 43,660/37 (C.P. XIV).
28. George to Andrew Combe, 26 January 1843. NLS 7380/5.
29. Cobden to Simpson, 4 July 1846. B.M. 43,660/52 (C.P. XIV).
30. Morley, *op.cit.*, I, p.203.
31. Cobden to Combe, 15 May 1848. B.M. 43,660/116. (C.P. XIV).
32. Cobden to Combe, 24 April 1848. B.M. 43,660/113 (C.P. XIV).
33. Cobden to Combe, 15 May 1848. B.M. 43/660/117 (C.P. XIV).
34. Morley, *Life of Cobden*, II, p.6.
35. Cobden to Combe, 24 April 1848, *op. cit.*
36. Cobden to Combe, 5 January 1849. B.M. 43,660/175. (C.P. XIV).
37. Combe to Cobden, 13 January 1848. B.M. 43,660/93 (C.P. XIV).
38. Cobden to Combe, 18 November 1850. B.M. 43,660/251 (C.P. XIV).
39. Cobden to Combe, 9 November 1850. B.M. 43,660/247. (C.P. XIV).
40. Mainzer to Combe, 14 May 1849. NLS 7303/3.
41. *Westminster Review*, n.s. II (July 1852), p.31.
42. *Hansard* [3rd ser.] CX (1850), c.475–78.
43. *ibid.*, CIX, c.1239.
44. Fox to Combe, 5 May 1850. NLS 7308/15.
45. Among them was Joseph Hume, member for Montrose.
46. *Hansard* [3rd ser.] CXXV (1853), c.533–34.
47. Cobden to Combe, 5 January 1849, B.M. 43,660/173 (C.P. XIV);

Lucas to Combe, 28 January 1854, NLS 7342/71; Hume to Combe, 4 May 1847, NLS 7285/116; R.L. Archer. *Secondary Education in the Nineteenth Century* (Cambridge, 1921), p.120
48. Ellis to Combe, 22 December 1850. NLS 7307/144.
49. Cobden's compliment to Combe, 10 January 1848. B.M. 43,660/89 (C.P. XIV).
50. Combe to Lucas, 14 May 1848. NLS 7391/443.
51. Simpson, *Necessity of Popular Education*, p.213.
52. Watson to Combe, 21 March 1848, NLS 7298/44–47; 29 September 1848, NLS 7298/81.
53. Combe to Watson, 6 May 1848. NLS 7391/430.
54. Writing to Lucas on 7 June 1847 (NLS 7391/93), Combe expressed his 'great satisfaction' with the draft.
55. Lucas to Combe, 25 May 1849. NLS 7302/68.
56. Lucas to Combe, 13 July 1848. NLS 7294/80. A handbill in the Combe Papers (NLS 7286/64), explains the Association's plans for secular education as they were presented to parliament in 1850.
57. Smiles to Combe, 13 March 1850. NLS 7311/39.
58. Russell to Combe, 24 November 1849. NLS 7303/97.
59. Combe to Hodgson, 9 November 1850. NLS 7392/222.
60. Combe to Cobden, 27 April 1851. B.M. 43,661/9 (C.P. XV).
61. Edward Lombe to Combe, 17 December 1850. NLS 7309/93.
62. Lucas to Combe, 15 November 1851. NLS 7317/104.
63. Hickson to Combe, 18 May 1850. NLS 7308/78.
64. Combe to McClelland, 20 July 1851. NLS copy book.
65. Hickson to Combe, 3 October 1850. NLS 7308/84.
66. Smiles to Combe, 30 September 1851, NLS 7321/15; Lucas to Combe, 27 September 1851, NLS 7317/102; Lombe to Combe, 15 October 1851, NLS 7317/58.
67. Lombe to Combe, 12 July 1850. NLS 7309/91.
68. Combe to Lombe, 22 December 1851. NLS 7392/493.
69. Simpson, *Philosophy of Education*, p.140; *Report of the Speeches delivered at a Dinner*, p.15.
70. *Hansard* [3rd ser.], CX (1850), c.438–39.
71. A handbill of the National Public School Association, dated 8 January 1853. NLS 7336/2.
72. Ellis to Combe, 4 November 1851. NLS 7315/117.
73. Lovett, *Life and Struggles*, p.363.
74. Combe to Lovett, 18 October 1848. NLS 7391/555.
75. Lovett to Combe, 29 September 1849. NLS 7302/44.
76. Lovett to Combe, 22 November 1849. NLS 7302/46–47.
77. *ibid*.
78. Ellis to Combe, 29 April 1851. NLS 7315/74.
79. Andrew to George Combe, 29 May 1838. NLS 7246/35.
80. George to Andrew Combe, 2 May 1846. NLS 7381/12.
81. Elizabeth Longford *Victoria R.I.* (London, 1964), p.217.
82. George to Andrew Combe, 11 May and 12 June 1846. NLS 7381/12 and 36.
83. Andrew to George Combe, 20 June 1846. NLS 7278/157–62.

84. Andrew to George Combe, 9 September 1846. NLS 7278/190.
85. Combe to Clark, 27 November 1850. NLS 7392/249.
86. Combe to Clark, 20 June 1856. NLS 7353/6.

Afterword

Late in 1851, William Ellis wondered whether phrenologists had been wise to commit themselves to what he called the 'monstrous project' of national education. As he complained in a letter to George Combe, the extraparliamentary campaign had become a great physical and financial strain, and for all their efforts it now seemed that the members of the National Public School Association were about to be left empty-handed.[1] In later years Ellis was to reach the same conclusion about the 'palace project' which, for all its early promise, was as unproductive as parliamentary and popular agitation. Naturally Ellis and Dr Hodgson were bitter when they blamed the 'unenlightened and priest-ridden people' for the failure of national education in the 1850s; but what were they now to do? Having no popular and national figure who would crusade for education as long and as stubbornly as Cobden had for free trade, their only alternative was to do what *The Times* suggested in January 1852. Commenting on the mission of the Public School Association to Lord John Russell, *The Times* advised the Manchester men 'to do without legislation'; there was no need for anyone to reform education root-and-branch but (according to *The Times*) reformers should feel free to continue their experiments in model schools.[2] The advice amounted to a policy of educational co-existence, by which progressive schools might continue to spring up as rivals to the old grammar and classical schools. Certainly the rivalry had been fierce enough for the last two or three decades; in some cases new secular schools were built literally next door to the old parish institutions. Indeed, the proliferation of new schools had occurred at such a pace that reformers miscalculated the political strength of their cause.

The politics of education had been a difficult game to play, and the phrenologists (in good company with other reformers) had bargained for disappointment. The resignation, however, with which George Combe accepted the outcome, and the relief with which Hodgson, Ellis and Williams returned to their schools, should not imply that phrenologists had played the game badly. True, they did not instigate every working-class demonstration which occurred on behalf of secular education in 1851, and they were not the moving spirit behind Fox's bill of February 1850. The Manchester Public School Association was not fashioned as a tool for phrenological education, and the public never applauded the 'moral training' which phrenologist tutors gave to the royal children. But that is not to say that phrenologists were on the periphery of the educational movement. On the contrary, they were of critical importance in the founding of model schools, in the direction of the Public School Association, and in rousing the conscience of the country. They may have been amateurs at extracting promises from politicians, but the impression which they left before the select committee, through the testimony of Simpson and Dorsay, was a forceful one, and a contribution towards the ultimate success of national education. Phrenologists were too realistic to expect a complete victory to come of the agitation of 1851, which they considered only the first inning of a long contest. As William Ellis envisaged it, the real goal of phrenologists in the period from 1847 to 1852 was to help to keep the education question at a very high temperature and eventually, by popular pressure, parliamentary concession or royal prodding, to bring the issue to a boil.[3] In this capacity phrenologists made themselves useful. And ever since those days of model schools and mass meetings, Simpson, Ellis, Williams and Dorsay — if they are remembered at all — are associated with educational reform; their reputations as phrenologists have barely survived.

The same must be said, although perhaps with less precision, of those who applied phrenology to other causes of social reform. Archbishop Whately is known for his promotion of political economy as a school discipline and not for his phrenology or his 'somnabulic experiments'. Robert Macnish may possibly be listed among the more widely-read of the temperance writers, but his phrenology is forgotten. The same is true of Richard Carmichael and John Conolly, known for their plans of medical and hospital reform. Captain Maconochie is hardly remembered either as a penal reformer or as a phrenologist; the phrenology of leading Victorians has been cited either as evidence of their intellectual wanderlust or their attachment to anthropology. Others gave credit where credit was due, and admitted that the study of character was 'considerably awakened' by phrenology, whose supporters were on occasion able to 'marshall an array of facts . . . so formidable and cogent as almost to silence opposition'.[4] If in so many individual careers phrenology served as an intellectual parenthesis — if Samuel Gridley Howe, Hewett Watson and the Reverend David Welsh were all reluctant after a while to identify with the new science — they nonetheless continued to believe in intrinsic qualities of mind which distinguished men and nations. To this extent, phrenology was quietly maintained until very recent times; and in its more popular form, which Caldwell once dismissed as 'head-reading', it has survived as a rough guide to character and individual ability. When latterday phrenologists discussed George Combe's role in the education campaigns of the 1840s and 1850s, they were more aware of the weakness of phrenology as a mental science; but they were also satisfied that as a philosophy it had pointed the way to recent legislation in prisons, hospitals and schools.[5] As long as man stood in need of reformation, his Victorian reformers had to know where the process began and how far it might be carried, and many of them found their answers in phrenology.

Notes

1. Ellis to Combe, 15 December 1851. NLS 7315/134.
2. *The Times*, 8 January 1852.
3. Ellis to Combe, 14 October 1851. NLS 7315/103–04.
4. Alexander Bain, *On the Study of Character, including an Estimate of Phrenology* (London, 1861), p.16.
5. William Jolly, an inspector of schools, wrote his biography of Combe in 1877, seven years after the Forster Education Act.

Bibliography

I. Unpublished Materials

(a) Manuscripts
1. In the National Library of Scotland, Edinburgh:
    The George Combe Papers (NLS 7201–7515), including letters sent and received by George Combe, as well as his lecture notes, journals, financial papers and notebooks. Presented to the National Library in 1950 by the Combe Trustees.
2. In the British Museum, London:
    The Richard Cobden Papers.
    Two volumes of this collection (B.M. 43,660 and 43,661) cover almost twenty years of correspondence between Cobden and Combe.
3. In the Society for Promoting Christian Knowledge, London:
    Reports and Minutes of the Committee of General Literature and Education (1833).

(b) Other
Alastair Cameron Grant. 'George Combe and His Circle: with particular Reference to his Relations with the United States'. A dissertation for the Ph.D. degree at Edinburgh University, 1960.

II. The Writings of George Combe.

This list includes tracts and pamphlets, together with Combe's major works. Only the principal editions are given.

An Address delivered at the Anniversary Celebration of the Birth of Spurzheim, and the Organization of the Boston Phrenological Society, December 31, 1839. Boston, 1840.

Answer by George Combe to the Attack on the 'Constitution of Man', contained in 'Nature and Revelation harmonious... by the Rev. C. J. Kennedy'... Edinburgh, 1848.

The Constitution of Man considered in Relation to External Objects. Edinburgh, 1828.
  [1st American ed.] Boston, 1829.
  [2nd ed., corrected and enlarged] Edinburgh, 1835.
  ['People's Edition'] Edinburgh, 1836.
  [5th ed.] Edinburgh, 1835.
  [8th ed.] Edinburgh, 1847 [reprinted by Gregg International in 1970].
  [Another ed.] Boston, 1841.

[Another ed.] *The Constitution of Man, considered in Relation to the Natural Laws. Adapted for the Use of Schools.* New York, 1851.
[9th ed., revised by Robert Cox and James Coxe] Edinburgh, 1860.
The Currency Question, considered in Relation to the Act of the 7th and 8th Victoria, Chap. 32, commonly called the Bank Restriction Act... London, 1856.
[Most of this work originally appeared in the *Scotsman*, in November and December 1855]
Elements of Phrenology. Edinburgh, 1824.
[4th ed.] Edinburgh, 1836.
Essays on Phrenology: or an Inquiry into the Principles and Utility of the System of Drs. Gall and Spurzheim, and into the Objections made against it. Edinburgh, 1819.
[Later editions appeared as: *A System of Phrenology; or an Inquiry*...]
[2nd ed.] Edinburgh, 1825.
[3rd ed.] Edinburgh, 1830.
[4th ed., in 2 vols] Edinburgh, 1836.
[5th ed., in 2 vols] Edinburgh, 1843.
Lectures on Phrenology... including its Application to the present and prospective Condition of the United States... London, 1839.
Lectures on Popular Education; delivered to the Edinburgh Philosophical Association, in April and November 1833. Edinburgh, 1833.
[2nd ed.] Edinburgh, 1837.
[3rd ed.] Edinburgh and London, 1848.
Letter on the Prejudices of the Great in Science and Philosophy against Phrenology; addressed to the Editor of the Edinburgh Weekly Journal. Edinburgh, 1829.
Letter to Francis Jeffrey, Esq., in Answer to his Criticism on Phrenology, contained in No. LXXXVIII of the Edinburgh Review. London, 1826.
The Life and Correspondence of Andrew Combe, M.D. Edinburgh, 1850.
The Life and dying Testimony of Abram Combe in favour of Robert Owen's New Views of Man and Society. London, 1844.
Moral Philosophy; or, the Duties of Man considered in his individual, social, and domestic Capacities. Edinburgh, 1840.
[2nd ed.] Edinburgh, 1841.
[People's ed.] Edinburgh, 1846.
Notes in Answer to Mr. Scott's Remarks on Mr. Combe's Essay on the Natural Constitution of Man. Edinburgh, 1827.
Notes on a Visit to Germany in 1854. Edinburgh, 1854.
Notes on the New Reformation in Germany, and on National Education, and the Common Schools of Massachusetts. Edinburgh, 1845.
Notes on the United States of North America during a phrenological Visit in 1838, 39, 40. 3 vols. Edinburgh, 1841.
[America ed., in 2 vols.] Philadelphia, 1841.

On Human Responsibility as affected by Phrenology. Edinburgh, 1826. [This short essay was published privately for the use of the Phrenological Society of Edinburgh].

On Teaching Physiology and its Application in Common Schools. Edinburgh, 1857.

On the Relation between Religion and Science. Edinburgh, 1847. [also appears as chapter IX in the 1847 ed. of the *Constitution of Man*] [4th ed.] *On the Relation between Science and Religion.* Edinburgh, 1857.

Our Rule In India. Edinburgh, 1858 [reprinted from The *Scotsman*]

Phrenology applied to Painting and Sculpture. London, 1855.

Refutations refuted. A Reply to Pamphlets put forth in Answer to 'The Currency Question Considered'. London, 1856.

Remarks on National Education, being an Inquiry into the Right and Duty of Government to educate the People. Edinburgh, 1847.

Remarks on the Principles of Criminal Legislation, and the Practice of Prison Discipline. London, 1854.

Secular Education. Lecture on the comparative Influence of the Natural Sciences and the 'Shorter Catechism', on the Civilization of Scotland... Edinburgh, 1851.

Secular Instruction or Extention of Church Endowments? Letter to His Grace the Duke of Argyle...on the Church of Scotland's Endowment Scheme. Edinburgh, 1852. [originally appeared in the *Scotsman* during January 1852]

Thoughts on Capital Punishment. Edinburgh, 1847.

What Should Secular Education Embrace? Edinburgh, 1848.

[, in a review of] *Education as a Means of preventing Destitution; with Exemplifications from the Teaching of the Conditions of Well-being...by William Ellis,* in *Westminster Review,* n.s.II (July, 1852), 1–32.

         and C. J. A. Mittermaier. *On the Application of Phrenology to Criminal Legislation and Prison Discipline.* [London, 1843; printed in the *Phrenological Journal,* XVI, no.74 (1843)].

'George Combe, Robert Cox, and Others'. *Moral and Intellectual Science, applied to the Elevation of Society.* New York, 1848 [largely composed of Combe's *Moral Philosophy*].

III. Contemporary Works

Abercrombie, Dr. John. *Culture and Discipline of the Mind.* Edinburgh, 1837 [6th ed.]

An Address to the People of Scotland, occasioned by the present Disputes on the 'Constitution of Man', as These relate to scriptural Instruction and coercise Provisions for its Diffusion among the Population... By One of Themselves. Edinburgh, 1836.

[anon.] *Man: as a physical, moral, religious and intellectual Being, considered phrenologically.* Glasgow, 1844.

[anon.] *National Education: the present State of the Question elucidated, in some Remarks on the Plans submitted to Parliament at*

the last Session ... London, 1839.
[anon.] *Observations on Phrenology, as affording a systematic View of Human Nature.* London, 1822.
[anon.] *Orthodox Phrenology* London [n.d.]
[anon.] *The Phrenologist's Daughter: a Tale.* London, 1854.
[anon.] *Phrenology* London [1850?] [Three short essays, bound with 'The New Age, Concordium Gazette, and Temperance Advocate' and an essay on 'Electro-Magneticism', in the British Museum].
[anon.] *Phrenology, and the moral Influence of Phrenology.* London [1835].
[anon.] *Phrenology; in relation to the Novel, the Criticism, and the Drama.* London [1848].
[anon.] *Phrenology made easy; or the Art of Studying Character in relation to Love, Courtship, and Marriage, showing the best Means of cultivating Character, so as to arrive at early Wealth, Position and Happiness* ... London, [1870?].
Bain, Alexander. *On the Study of Character, including an Estimate of Phrenology* London, 1861.
Barnes, Dr Melvin. *A few general and unmethodized Remarks to a medical younger Friend, on Phrenology.* Plattsburgh, N.Y., 1853.
Bell, Charles. *The Nervous System of the Human Body.* London, 1830.
Braid, James. *The Power of the Mind over the Body: an experimental Enquiry into the Nature and Cause of the Phenomena attributed by Baron Reichenbach and Others to a 'New Imponderable'.* London, 1846.
Bray, Charles. *The Education of the Feelings. A System of Moral Training, for the Guidance of Teachers, Parents and Guardians of the Young.* London, 1838.
────── *The Philosophy of Necessity; or, Natural Law as applicable to moral, mental and social Science.* 2 vols. London, 1841.
[2nd ed.] London, 1863.
Butler, Bishop Joseph. *Analogy of Religion, natural and revealed, to the Constitution and Course of Nature.* Edinburgh, 1804.
Byam, W. *Explanation of the new physiognomical System of the Brain, according to Drs Gall and Spurzheim.* London, 1818.
Caldwell, Dr Charles. *Elements of Phrenology.* Lexington, Ky., 1824.
────── *Thoughts on Physical Education, and the true Mode of improving the Condition of Man; and on the Study of the Greek and Latin Languages,* ed. Robert Cox. Edinburgh, 1836.
Capen, Nahum. *Reminiscences of Dr Spurzheim and George Combe* ... Boston, 1881.
Carmichael, Dr Richard. *Plan of Medical Reform ... an Address to the Rt. Hon. Sir Robert Peel.* Dublin, 1841.
Central Society of Education. *Schools for the Industrious Classes; or, the present State of Education among the Working People of England.* London, 1837 [2nd ed.]
Challenge to Phrenologists; or, Phrenology tested by Reason and Facts, by A. M. of Middle Temple. London, [1839?].
[Chambers, Robert]. *The Vestiges of the Natural History of Creation.*

London, 1844 [2nd ed.]
Changes produced in the Nervous System by Civilizations, considered according to the Evidence of Physiology and the Philosphy of History. London, 1839.
Chenevix, Richard. 'Phrenology', in the *Foreign Quarterly Review.* London, 1830.
Church, Richard. *Presumptive Evidence of the Truth and Reasonableness of Phrenology . . .* Chichester, 1833.
Clark, Dr Sir James. *A Memoir of John Conolly, M.D.* London, 1869.
—— *Remarks on Medical Reform; in a second Letter addressed to the Rt Hon. Sir James Graham.* London, 1843.
Clarke, Henry. *Christian Phrenology; or, the Teachings of the New Testament respecting the animal, moral, and intellectual Nature of Man. Three Lectures delivered in the Thistle Hall, Dundee . . . to which are added a Phrenological Table . . .* Dundee, 1835.
[Coleridge, S.T.] *The Table Talk and Omniana of Samuel Taylor Coleridge,* ed. T. Ashe. London, 1923 [reprint].
Colquhoun, John C. *The Uses of the Established Church to the Protestantism and Civilization of Ireland.* London, 1836.
Combe, Abram. *An Address to the Conductors of the Periodical Press, upon the Causes of religious and political Disputes . . .* Edinburgh, 1823.
Combe, Dr Andrew. *Observations on Mental Derangement . . .* Edinburgh, 1830.
—————— *On the Introduction of Religion into Common Schools.* London and Edinburgh, 1850 [extracts edited by George Combe]
—————— *Phrenology – its Nature and Uses: an Address to the Students of Anderson's University.* Edinburgh, 1846.
—————— *Phrenology, philosophically and physiologically considered: with Reasons for its Study, and Directions for its successful Prosecution.* London, 1840.
—————— *The Principles of Physiology, applied to the Preservation of Health, and to the Improvement of Physical and Mental Education.* Edinburgh, 1835. [3rd ed.]
[5th ed.] Edinburgh, 1836.
—————— *A Treatise on the physiological and moral Management of Infancy: being a practical Exposition of the Principles of Infant Training, for the Use of Parents.* Edinburgh, 1840.
Conolly, Dr John. *A Inquiry concerning the Indications of Insanity, with Suggestions for the better Protection and Care of the Insane.* London, 1830.
[reprint: Psychiatric Monograph Series, London, 1964].
Cox, Robert. *Sabbath Laws and Sabbath Duties considered in relation to their natural and scriptural Grounds . . .* Edinburgh and London, 1853.
——, ed. *The Literature of the Sabbath Question.* 2 vol. Edinburgh, 1865.
Darwin, Francis, ed. *The Life and Letters of Charles Darwin. Including an autobiographical Chapter.* 3 vols. London, 1887.

Bibliography 235

Discussion on Phrenology, between Charles Donovan and the Reverend Brewin Grant. London, 1850.
Documents laid before the Rt Hon. Lord Glenelg, by Sir George Mackenzie, relative to the Convicts sent to New South Wales. April, 1836 [Edinburgh] 1836.
Edinburgh Essays, by Members of the University. Edinburgh, 1856 [nos. 3 and 7].
Editor of the Scottish Guardian. *Scotland: a half-educated Nation, both in Quantity and Quality of her educational Institutions*. Glasgow, 1834.
[George Eliot (Marian Evans)] *George Eliot's Life as related in her Letters and Journals*, ed. J. W. Cross. 3 vols. London, 1885.
Ellis, William. *An Address to Teachers on the Laws of Conduct in Industrial Life, and on the Method of imparting Instruction therein in our Primary Schools* . . . London [1870].
―――― *Education as a Means of preventing Destitution: with Exemplifications from the Teaching of the Condition of Well-being and the Principles and Applications of Economical Science, at the Birkbeck Schools*. London, 1851.
―――― *Thoughts on the Future of the Human Race*. London, 1866.
―――― *What am I? Where am I? What ought I to do?* . . . London, 1852.
[――――] 'Classical Education', in the *Westminster and Foreign Quarterly Review*, LIII (July 1850).
Engledue, Dr W. C. *Cerebral Physiology and Materialism . . . an Address . . . with a letter from Dr Elliotson*. London, 1842.
―――― *Some Account of Phrenology, its Nature, Principles and Uses*. Chichester, 1837.
Epps, Dr John. *Horae Phrenologicae: being three phrenological Essays* . . . London, 1829.
[2nd ed.] London, 1834.
―――― *Internal Evidences of Christianity, deduced from Phrenology*. Edinburgh, 1835.
[――――] *Diary of the late John Epps, M.D.*, ed. Ellen Epps. London [1875].
Gall, Dr F. J. *Sur les fonctions du cerveau et sur celles de chacune de ses parties, avec des observations sur la possibilité de reconnaître les instincts, les penchans, les talons, ou les dispositions morales et intellectuelles des hommes et des animaux* . . . 6 vols. Paris, 1825 [another ed., trans. by Winslow Lewis, Jr.] Boston, 1835.
――――, and Johann Spurzheim. *Recherches sur le Système nerveux en général et sur celui du cerveau en particulier*. Paris, 1809–11.
Gillespie, William. *Exposure of the Unchristian and Unphilosophical Principles set forth in Mr George Combe's Work . . . being an Antidote to the Poison of that Publication*. Edinburgh, 1836.
Glasgow Educational Society. *Hints towards the Formation of a Normal Seminary in Glasgow, for the professional Training of School-masters* [Glasgow, 1835; includes letters by Rev. David Welsh].

Gordon, Dr John. *Observations on the Structure of the Brain, comprising an Estimate of the Claims of Drs Gall and Spurzheim, to the Discovery of the Anatomy of that Organ* ... [Edinburgh] 1817.
Goyder, D. G. *Acquisitiveness, its Uses and Abuses.* Edinburgh, 1836.
─────── *Autobiography of a Phrenologist.* London, 1857.
─────── *Treatise on the Management of Infant Schools.* London, 1826.
Guthrie, Dr G. J. *A Letter to the Rt Hon. the Secretary of State for the Home Department, containing Remarks on the Report of the Select Committee of the House of Commons, on Anatomy* ... London, 1829.
Halliwell, Thomas. *Examination and Refutation of Phrenology.* Dunedin, 1864.
Hodgson, Rev. J. S. *Considerations on Phrenology: in Connexion with an intellectual, moral and religious Education.* London, 1839.
Hodgson, Dr William B. *Address delivered to the Mental Improvement Society of the Liverpool Mechanics' Institution.* Liverpool [1845].
─────── *Second Annual Address delivered to the Mental Improvement Society* ... Liverpool, 1846.
Holyoake, George Jacob. *English Secularism: A Confession of Belief* [American ed.] Chicago, 1896.
─────── *Sixty Years of an Agitator's Life.* 2 vols. London, 1892.
Huxley, Thomas. *Science and Education.* New York, 1902.
Jolly, William. *George Combe as an Educationalist.* [London, 1877] [another ed., enlarged] London, 1879.
Jones, Philip. *Popular Phrenology tried by the Word of God, and proved to be Anti-Christ, and injurious to Individuals and Families.* London, 1845.
Kemble, Frances Ann. *Records of Later Life.* 3 vols. London, 1882.
Kennedy, Rev. C. J. *Nature and Revelation harmonious; a Defence of Scriptural Truths assailed in Mr George Combe's Work on 'The Constitution of Man'.* Edinburgh, 1846.
Knox, Vicesimus. *Remarks on the Tendency of certain Clauses in a Bill now in Parliament to degrade the Grammar Schools* ... London, 1821.
Levison, J. L. *Lecture on the hereditary Tendency of Drunkenness.* London, 1839.
─────── *Mental Culture; or, the Means of Developing the Human Faculties.* London, 1833.
Lovett, William. *Life and Struggles of William Lovett in his Pursuit of Bread, Knowledge and Freedom* ... London, 1878.
Mackenzie, Sir George S. *General Observations on the Principles of Education: for the Use of Mechanics' Institutions.* Edinburgh, 1836.
─────── *Representation to the Rt Hon. Lord Glenelg* ... Edinburgh, 1836.
─────── *Illustrations of Phrenology.* Edinburgh 1820.
─────── *Three Lectures on the Insufficiency of physical Facts for establishing the continued Existence of the Deity; and on the Superiority of the Proofs that may be derived from the Structure of the Human Mind, and its Adaptation to the external World.* London

[1840?]
Macnab, Henry G. *Analysis and Analogy recommended as the Means of rendering Experience and Observation useful in Education*. Paris, 1818.
Macnish, Robert. *The Anatomy of Drunkenness*. Glasgow, 1827.
[another ed.] New York, 1835.
―――― *An Introduction to Phrenology, in the Form of Question and Answer*. Glasgow, 1836.
―――― *The Philosophy of Sleep*. Glasgow, 1830.
[another ed.] New York, 1835.
Maconochie, Alexander. *Comparison between Mr Bentham's Views on Punishment, and Those advocated in Connexion with the Mark System*. London [1847].
―――― *Crime and Punishment*. London, 1846 [bound and paginated with:
The Principles of the Mark System and *Secondary Punishment*]
―――― *Norfolk Island*. London, 1847.
――――, ed. *Copy of a Dispatch from the Lieutenant-Governor Sir John Franklin to Lord Glenelg, dated 7th October, 1837, relative to the present System of Convict Discipline in Van Diemen's Land*. Hobart Town [Tasmania], 1838.
Manual for Mechanics' Institutions, ed. B. F. Duppa. London, 1839.
Martineau, Harriet. *Biographical Sketches*. London, 1869.
Mège, J. B. *Des Principes fondamentaux de la Phrénologie appliqués à la Philosophie*. Paris, 1845.
Member of the City Philosophical Society. *Three familiar Lectures on craniological Physiognomy*. London, 1816.
A Member of the Phrenological Society. *A Key to Phrenology, containing a brief Statement of the Faculties of the Mind, the History and practical Uses of Phrenology*. Edinburgh [1836]
Nicol, John P. Life of George Combe, *Imperial Dictionary of Universal Biography*, VI, p.1100. Glasgow [1880]
Noble, Dr Daniel. *Essays on the Means, physical and moral, of estimating Human Character*. Manchester, 1835.
Noel, Robert. *Die Begründung und das Wesen der Phrenologie*. Dresden, 1852.
'On Phrenology', in the *Quarterly Review*, no.113 (September 1836).
'Phrenology', in *Penny Cyclopedia*. London, 1833.
Porter, G. R. *Progress of the Nation*. London, 1843.
Prideaux, Thomas Symes. *Gall's Organology ... reprinted from the Anthropological Review*. London, [1869].
―――― *Strictures on the Conduct of Hewett Watson, in his Capacity as Editor of the Phrenological Journal* London, 1840.
Recent Measures for the Promotion of Education in England. London, 1839. [4th ed.]
Reid, Thomas. *Essays on the Intellectual Powers of Man*. Edinburgh, 1785.
Rennell, Rev. Thomas. *Remarks on Scepticism, especially as it is connected with the Subjects of Organization and Life . . .* London,

1819.

Sampson, M. B. *Criminal Jurisprudence considered in relation to mental Organization*. London, 1841.

───── *The phrenological Theory of the Treatment of Criminals defended, in a Letter to John Forbes, M.D., Editor of the British and Foreign Medical Review*. London, 1843.

───── *Rationale of Crime and its appropriate Treatment*... New York, 1846.

Saunders, Sir Edwin. *What is Phrenology? Its Evidence and Principles familiarly considered*. London, 1835 [2nd ed.]

Scott, William. *The Harmony of Phrenology with Scripture: shewn in a Refutation of the philosophical Errors contained in Mr Combe's 'Constitution of Man'* Edinburgh, 1836.

Scoutetten, Henri. *Eléments de Philosophie phrénologie*. Metz, 1861.

Seedair, Stephen. *A Tract for all Time. The Christian or true Constitution of Man, versus the pernicious Fallacies of Mr Combe and other materialistic Writers*. Edinburgh, 1856.

Simpson, James. *Brief Reports of Lectures delivered to the Working Class of Edinburgh, on the Means in their own Power of improving their Character and Condition*. Edinburgh, 1844.

───── *Hints on the Formation and Conduct of a General Model Normal School, for Training Teachers to supply the Demand of a National System of Popular Education*. Edinburgh [n.d.; also appears as appendix VI to the *Philosophy of Education*, q.v.]

───── *The Necessity of Popular Education, as a national Object; with Hints on the Treatment of Criminals and Observations on Homicidal Insanity*. Edinburgh and London, 1834.

───── *The Normal School as it ought to be: its Principles, Objects, and Organisation*. Edinburgh, 1850.

───── *The Philosophy of Education, with its practical Application to a System and Plan of Popular Education as a national Object*. Edinburgh, 1836 [2nd ed.]

─────, ed. *Anti-National Education; or, the Spirit of Sectarianism morally tested by Means of certain Speeches and Letters from the Member for Kilmarnock*. Edinburgh, 1837.

Slade, Dr John. *Letters on Phrenology*. London [1836?].

Smith, Joshua Toulmin. *Local Self-Government and Centralization: the Characteristics of Each and its practical Tendencies as effecting moral, social, and political Welfare and Progress*... London, 1851.

───── *The Reasonableness of Phrenology; containing a Sketch of the Origin, Progress, Principles, Proofs and Tendencies of that Science*. London, 1837.

───── *Synopsis of Phrenology: directed chiefly to the Exhibition of the Utility of the Science to the Advancement of social Happiness*. London [n.d.].

Smith, Sidney. *The Principles of Phrenology*. Edinburgh, 1838.

Solly, Samuel. *The Human Brain, its Configuration, Structure, Development, and Physiology; illustrated by References to the Nervous System in the Lower Order of Animals*. London [1835?]

Spencer, Herbert. *Autobiography.* 2 vols. London, 1904.
Spurzheim, Dr Johann. *Essai philosophique sur la Nature morale et intellectuelle de l'Homme.* Paris, 1820.
―――― *Observations sur la Follie, ou sur les Dérangemens des Fonctions morales et intellectuelles de l'Homme.* Paris, 1817.
―――― *Observations sur la Phrénologie, ou la Connaissance de l'Homme moral et intellectuel, fondée sur les Fonctions du Système nerveux.* Paris, 1818.
―――― *Philosophical Cathechism of the Natural Laws of Man.* London, 1826.
[another ed.] Boston, 1835.
―――― *The physiognomical System of Drs Gall and Spurzheim; founded on an anatomical and physiological Examination of the Nervous System in general, and of the Brain in particular, and indicating the Dispositions and Manifestations of the Mind...* London, 1815.
―――― *A View of the elementary Principles of Education, founded on the Study of the Nature of Man.* London and Edinburgh, 1821.
―――― *View of the Philosophical Principles of Phrenology.* London [3rd ed., 1845].
Testimonials on Behalf of George Combe, as a Candidate for the Chair of Logic in the University of Edinburgh. Edinburgh, 1836.
Turner, Henry. *Phrenology: its Evidences and Inferences, with Criticisms upon Mr Grant's recent Lectures.* London and Sheffield, 1858 [2nd ed.]
Tyrrell, Frederick. *An Introductory Lecture on Anatomy; delivered at the New Medical School, Aldersgate Street...* London, 1826.
Verity, Robert. *Changes produced in the nervous System by Civilization...* London [1839].
Vimont, Dr Joseph. *Traité de Phrénologie humaine et comparée.* 2 vols. Paris and London, 1835.
Wallace, Alfred Russel. *The Wonderful Century, its Successes and its Failures.* London, 1898.
Warne, Rev. Joseph. *On the Harmony between the Scriptures and Phrenology.* Edinburgh, 1836.
Watson, Hewett C. *An Examination of Mr Scott's Attack upon Mr Combe's 'Constitution of Man'.* London, 1836.
―――― *Statistics of Phrenology; being a Sketch of the Progress and present State of that Science in the British Islands.* London, 1836.
Watson, Robert S. *The History of the Literary and Philosophical Society of Newcastle-upon-Tyne.* London, 1897.
Whately, Richard. *Elements of Rhetoric.* London, 1836 [5th ed.]
―――― *Reply of His Grace the Archbishop of Dublin to the Address of the Clergy of the Dioceses of Dublin and Glandelough, on the Government Plan for National Education in Ireland...* London, 1832 [2nd ed.]
―――― *Thoughts on Secondary Punishment, in a Letter to Earl Grey... to which are appended, Two Articles on Transportation to New South Wales, and on Secondary Punishments, and some*

Observations on Colonization. London, 1832.
Whately, E. Jane, ed. *Life and Correspondence of Richard Whately, D.D.* 2 vols. London, 1866.
Wilderspin, Samuel. *Importance of educating the Infant Poor...* London, 1824. [2nd ed.]
———— *A System for the Education of the Young, applied to all Faculties.* London, 1840.
Williams, William Mattieu. *A Vindication of Phrenology...* London, 1894.
———— *Who should teach Christianity to the Children? The Schoolmasters or the Clergy?* Edinburgh, 1853.
Wyse, Thomas. *Education Reform; or, the Necessity of a National System of Education.* London, 1836.
[————] *Speech delivered at the Opening of the New Mechanics' Institution, Mount Street, Liverpool, on the 15th of September, 1837...* Liverpool [1837].

IV Reports and Proceedings

(a) Official.
Hansard, 3rd series. 1846, 1850, 1852, 1853.
Report from the Select Committee on the State of Education in England and Wales. 1835.
Report from the Select Committee on Manchester and Salford Education; together with the Proceedings of the Committee, Minutes of Evidence... 1852.

(b) Private.
Report of a Committee of the Manchester Statistical Society, on the State of Education in the Borough of Bury, Lancashire, in July, 1835. London, 1835.
Report of a Committee of the Manchester Statistical Society, on the State of Education in the Borough of Salford in 1835. London, 1836.
Report of a Discussion regarding Ragged Schools; with the Speeches of Lord Murray, Sheriff Speirs, Prof. Gregory, Rev. Thomas Guthrie, James Simpson, Rev. Dr Alexander... held in the Music Hall, Edinburgh... Edinburgh, 1847.
Reports of the Edinburgh Secular School [compiled by W. M. Williams and published as an appendix in William Jolly's *George Combe as an Educationalist. First* through *Fourth Reports,* 1850–3]
Report of the Proceedings at a Meeting of the Working Classes of Edinburgh, on National Education, held in the Waterloo Rooms there on 21st January 1851. Edinburgh, 1851 [reprinted from the *Scotsman.*]
Report of the Proceedings at a Public Meeting held in the Mechanics' Hall. Aberdeen, April 25, 1851, on National Education. [n.p.] 1851
Report of the Proceedings of the Phrenological Society, since its Establishment... to the Close of the Second Session. Edinburgh

1821.
Report of the Speeches delivered at a Dinner given to James Simpson, Esq., by the Friends of Education in Manchester; Thomas Wyse, M.P., in the Chair. Manchester, 1836.
Summary of the Proceedings of the Edinburgh Philosophical Association, from its Institution in 1832 to June, 1836. Edinburgh, 1836.
[forms part of appendix IV of James Simpson's *Philosophy of Education*.]
Third Annual Report of the Sheffield Phrenological Society. Sheffield, 1845.
Transactions of the Phrenological Society of Edinburgh. Edinburgh, 1823.

V. Newspapers and Periodicals
(dates of publication given only for phrenological items)

The Aberdeen Herald.
The American Phrenological Journal, 1857.
The Caledonian Mercury.
The Chambers' Edinburgh Journal.
The Christian Physician and Anthropological Magazine [later becoming] *The Phrenological Magazine and Christian Physician* [ed. Dr John Epps] London, 1835–39.
The Dublin Journal of Medical Science.
The Edinburgh Review.
The Gentleman's Magazine.
The Illustrated London News.
Illustrations of Phrenology: comprising Accounts of the Lives of Persons remarkable in some mental Respect, whether of Intellect or Feeling, and accurate Delineations of their Heads. London, 1841 [ed., G. R. Lewis]
The Lancet.
The London Litarary Gazette.
The Medical Gazette.
The Medico-Chirurgical Review.
The Morning Chronicle.
The New Moral World.
The North British Daily Mail.
The Phrenological Almanac, or Physiological Annual. Glasgow (1841? – 1845).
The Phrenological Journal and Miscellany. Edinburgh, 1823–1837. [continued as] *The Phrenological Journal and Magazine of Moral Science,* London, 1837–1840; Edinburgh, 1841–1847.
The Phrenological Magazine: a Journal of Education and Mental Science. London, 1880–84.
The Phrenomagnet and Mirror of Nature. London, 1843.
The Quarterly Journal of Education (Society for the Diffusion of Useful Knowledge).

The Scotsman.
The Spectator.
The Times.
The Westminster Review. [continued, 1846–1851, as *Westminster and Foreign Quarterly Review*]
The Zoist: a Journal of Cerebral Physiology and Mesmerism, and their Application to Human Welfare, London, 1843–56.

VI. Secondary Works

(a) Books.

Ackerknecht, Erwin, and Henri Vallois. *Franz Joseph Gall, Inventor of Phrenology and his Collection*, trans. Claire St Leon. Madison, Wisconsin, 1956.

Archer, R. L. *Secondary Education in the Nineteenth Century.* Cambridge, 1921.

Barry, Sir John Vincent. *Alexander Maconochie of Norfolk Island. A Study of a Pioneer in Penal Reform.* Melbourne, 1958.

Blyth, E. K. *Life of William Ellis.* London, 1889.

Boring, Edwin G. *A History of Experimental Psychology.* New York. 1950.

Davies, John D. *Phrenology, Fad and Science. A Nineteenth Century American Crusade.* New Haven, 1955.

Flugel, J. C. *A Hundred Years of Psychology, 1833–1933: with additional Part on Developments, 1933–1947.* London, 1951.

Gibbon, Charles. *The Life of George Combe.* 2 vols. London, 1878.

Gillisipie, Charles Coulston. *Genesis and Geology: a Study in the Relations of scientific Thought, Natural Theology, and Social Opinion in Great Britain, 1790–1850.* New York, 1959 [3rd ed.]

Harrison, J. F. C. *Quest for the New Moral World: Robert Owen and the Owenites in Britain and America.* New York, 1969.

Hedderly, Frances. *Phrenology: a Study of Mind.* London, 1970.

Himmelfarb, Gertrude. *Darwin and the Darwinian Revolution.* London, 1959.

Holland, Bernard. *The Unknown Life and Works of Dr Francis Joseph Gall, the Discoverer of the Anatomy and Physiology of the Brain . . .* London [1909?].

Inglis, Brian. *Fringe Medicine.* London, 1964.

Knox, H. M. *Two Hundred and Fifty Years of Scottish Education, 1696–1946.* Edinburgh, 1953.

Lanteri-Laura, Georges. *Histoire de la phrénologie. L'homme et son cerveau selon F. J. Gall.* Paris, 1970.

Leigh, Denis. *The Historical Development of British Psychiatry.* vo.I: *Eighteenth and Nineteenth Centuries.* Edinburgh, 1961.

Longford, Elizabeth. *Victoria R.I.* London, 1964.

MacCabe, Joseph. *Life and Letters of George Jacob Holyoake.* 2 vols. London, 1908.

Maltby, S.E. *Manchester and the Movement for National Elementary Education, 1800–1870.* Manchester, 1918.

Meiklejohn, J. M. D. *Life and Letters of William Ballantyne Hodgson*. Edinburgh, 1883.
Morley, John. *Life of Richard Cobden*. 2 vols. London, 1881.
Simon, Brian. *Studies in the History of Education, 1780–1870*. London, 1960.
Stern, Madeleine. *Heads and Headlines: the Phrenological Fowlers*. University of Oklahoma, 1971.
Stewart, W. A. C., and W. P. McCann. *The Educational Innovators, 1750–1880*. London, 1967.
Tylecote, Mabel. *The Mechanics' Institutes of Lancashire and Yorkshire before 1851*. Manchester, 1957.
White, R. J. *Political Tracts of Wordsworth, Coleridge and Shelley*. Cambridge, 1953 [introduction]
Widdess, John David. *A History of the Royal College of Physicians of Ireland, 1654–1963*. Edinburgh, 1963.
Willey, Basil. *The Eighteenth Century Background*. London, 1962.
Williams, Harley. *Doctors Differ. Five Studies in Contrast*. London, 1947 [includes section on John Elliotson.]
Young, R. M. *Mind, Brain and Adaptation in the Nineteenth Century*. Oxford, 1970.

(b) Articles.
Ackerknecht, Erwin. 'Contributions of Gall and the Phrenologists to the Knowledge of Brain Function', *The Brain and its Functions*. ed. F. N. L. Poynter. Oxford, 1958.
Collins, Philip. 'When Morals lay in Lumps', *The Listener*, XC, no.2316 (August 1973), pp.213–15.
de Giustino, David. 'Reforming the Commonwealth of Thieves: British Phrenologists and Australia', *Victorian Studies*, XV, no.4 (1972), pp.439–61.
Garnett, R. G. 'E. T. Craig: Communitarian, Educator, Phrenologist', *Vocational Aspect of Secondary and Further Education*, XV, no.31 (1963), pp.135–50.
Goldstrom, J. 'Richard Whately and Political Economy in School Books, 1833–1880', *Irish Historical Studies*, XVI, no.58 (1966), pp.131–46.
Grant, Alastair C. 'Combe on Phrenology and Free Will: a Note on Nineteenth-century Secularism', *Journal of the History of Ideas*, XXVI, no.1 (1965), pp.141–47.
Knoepflmacher, U. C. 'The Use of Classification: "The Wellesley Index" ', *Victorian Studies*, X, no.3 (1967), pp.263–67.
Maconochie, K. J. 'Alexander Maconochie: Sociologist and Penal Reformer', *Howard Journal*, IX, no.3 (1956), pp.235–41.
Price, Alan. 'A Pioneer of Scientific Education, George Combe', *Educational Review* (Birmingham), XII, no.3 (1960), pp.219–29.
Riegel, Robert E. 'The Introduction of Phrenology to the United States', *American Historical Review*, XXXIX, no.1 (1933), pp.73–78.

Index

Abercrombie, James (Lord Dunfermline), 201, 205–6, 210
Abercrombie, Dr John, 61
Argyle, Duke of, 207

Bain, Alexander, 74–5
Becker, Ernest, 222–3
Bell, Sir Charles, 2, 3, 32
Bodichon, Barbara, 191
Bray, Charles: democratic ideas 65, 209; on human character 70; as a lecturer 87; a materialist 167; an Owenite 109, 140; a pessimist 90; and phreno-mesmerism 98–9
Brown, Samuel, 106, 222
Brougham, Lord Henry, 175, 199, 200, 207

Caldwell, Dr Charles: ambitious lecturer 88; anti-metaphysical 36; health and hygiene 73, 186; leading American phrenologist 25; on mental depravity 62; and phreno-mesmerism 47; on popular phrenology 89, 229
Carmichael, Dr Richard, 8, 46, 86, 109, 228
Cerebral localisation, theory of, 13, 43
Chambers, Robert, 51–2, 122, 209
Christian Phrenologists, 108, 119–126, 130
Clark, Dr Sir James: compliments Ellis 182; friend of Andrew Combe 45; education of royal children 220–22; medical training 40, 46
Clarke, Henry, 120
Cobden, Richard: caution on secular education 205–6, 215; conversion to phrenology 203–4; and diffusion of phrenology 87, 93, 188; education reformer 183, 204–8; and natural religion 126; on promoting Temperance 111
Coleridge, S.T., 1, 23, 74

Combe, Abram, 5, 9, 64, 142–3
Combe, Dr Andrew: on character formation 17, 144; on classroom conditions 189; on convict rehabilitation 146; his education 4; influence at court 200; a leading physiologist 44–5; on natural religion 112; and phreno-mesmerism 98; supports Temperance 62–3, 73
Combe, George: candidacy for university chair 7–8; and capital punishment 152; conversion to phrenology 9; distrust of Owenism 142–4, 169; his dogmatism 28, 100; education in Edinburgh 3–4; education reformer 191, 227; and evolutionary theory 50–4, 184; faith in British race 64; as a lecturer 87; and Maconochie 160–1; on moral education 166; natural religion and scepticism 34, 117, 125–133; as peacemaker and socialite 90; on penal policies 147–151; and phreno-mesmerism 47–8; reputed 'self-esteem' 6–7; respect for natural laws 9, 112; role in Australian petition 154; sponsors secular school 179, 213; style of writing 26–9; tour of America 25–6, 84; urges Cobden on education 204–7; view of industrial revolution 66; visits Buckingham Palace 220–1, 223
Combeists: on human progress 137; natural religion of 111, 126–133, 179; and secular education 206
Common Sense, School of, 36–9
Conolly, Dr John, 40 46–7, 228
Constitution of Man: and evolution 50–4; handbook of phrenology 78; influence on educators 183, 188, 213, 218; materialism of 65; natural history of 59; on natural morality 34, 113; natural religion of 124–5, 130, 180;

Index

popularity of 3, 7, 29, 60, 82, 86, 90, 92, 97; racialism of 71; rationalism of 37, 55; read at court 220; religious views of 108, 112, 117, 119, 121, 175; and science 49
Cox Robert, 4, 67–8, 100, 179
Craig, Edward T., 44, 140, 189

Darwin, Charles, 3, 6, 35, 51–3
Derangement and insanity, 47, 62–3, 105, 137–8
Deville, James: alliance with Elliotson 95, 97; 'Great Apostle of Phrenology' 94; and Prince Albert 220; Transportation System 155–6
Dorsay, A.J.: English teacher 181–2; and Glasgow phrenologists 93; teacher of phrenology 188; testimony to Select Committee 228

Economic liberalism, 65–6
Eliot, George (Marian Evans): attracted to phrenology 33; belief in natural theology 111, 130, 132; praise of George Combe 5
Elliotson, Dr John: attracted to phrenology 44–5; care of patients 46; and London phrenologists 94–5; and penology 152, 154; and phreno-mesmerism 48, 96, 98; upholds Gall 79, 95–6; and Prince Albert 220; and the *Zoist* 81, 98
Ellis, William: against Evangelicals 107; and William Lovett 217–8; practical education 127–9, 171, 180, 182–3; on problems of popular agitation 209, 216, 219, 228; promotes female education 72; religious indifference of 126, 130, 182; and social sciences 127, 183; supports Williams School 179; utilitarianism of 152; views on progress 64
English, teaching of, 181–2
Epps, Dr John: advised by phrenology 61; decries ritualism 107; equalitarianism of 65; explains morality 138; interest in homoeopathy 99; materialism of 167; natural morality 113; an Owenite 109; use of phreno-mesmerism 48; wants practical psychology 35, 39

Fox, W.J., 198, 207–8, 211, 215

Gall, Franz Josef: accused of materialism 14, 39, 114; career and research 12–3; differences with Spurzheim 16–7, 20, 97; reception in Paris 15; upheld by Elliotson 96, 99; and Spurzheim 3, 14–5, 35, 38, 40, 108, 113, 172
Gibson, Milner, 211, 216
Glenelg, Lord (Charles Grant), 154–6
Gospel of Work, 64–5, 68, 132

Hallyburton, D.G., 153
Hamilton, Sir William, 7, 41–2, 92
Hodgson, Dr William B: classless education 197; friend of Lovett 217–8; secular education 189, 212, 227
Holyoake, George Jacob, 28, 34, 128, 215
Hume, Joseph, 201

Infant schools, 174–8, 186, 190, 198–9
Ireland, 69, 198, 202

Jeffrey, Francis, 6, 28, 80, 90, 177, 201

Lancashire Public School Association (later National Public School Association), 210–19, 227–8
Lancasterian System, 189–90
Levison, J.L., 19, 61, 63, 68, 82
Lombe, Edward, 212–15
Lucas, Samuel, 203, 207–216, 219

Maconochie, Alexander: inspired by phrenology 178, 228; on Norfolk Island 157; psychological approach 158–9; silence on phrenology 159–61
Mackenzie, Sir George Stewart: career in science 49–51, 222; on Cobden 203; critical of Anglicans 107; and Elliotson 95, 98, favours national association 88; and Glasgow phrenologists 93; and

natural religion 126; on morality 167; and Transportation System 154
Maclaren, Charles, 67, 80, 154
Macnab, Henry G., 38, 167
Macnish, Dr Robert, 33, 43–4, 228
Mainzer, Joseph: career 184–5; education experiments 185; religious opinions 110, 206
Mann, Horace: association with George Combe 6, 26–6, 107; inspires Lucas 210; and McClaren 202; work in education 173
Martineau, Harriet: on Combe 5, 6, 9; on phrenology 29, 33; on phrenomesmerism 47, 99; on Whately 123
McClaren, Duncan, 202
Mechanics' Institutes: lectures by phrenologists 61, 87, 91, 181; meetings on Transportation System 154; read *Constitution of Man* 3, 60, 92; and social sciences 183; used as schools 189, 198
Medical profession: attentive to Spurzheim and phrenology 48–9; division over phrenology 40–3, 93; and phrenomesmerism 48; search for new ideas 46–7
Mental culture and discipline, 61, 171
Mesmer, Franz Anton, 2
Mesmerism, 96, 98
Mittermaier, C.J.A., 147, 150
Molesworth, Sir William, 156–7
Music, teaching of, 184

Natural laws: basis of true religion 104, 131, 138; in *Constitution of Man* 37; in economic life 183; guide to education 175; spur to human progress 54–5, 115, 168
Natural morality, 113, 150, 168
Natural theology, 112, 128–9

Origin of Species, (see Darwin)
Owen, Robert; and Bray 87; and Abram Combe 9; considered irreligious 105, 176; connections with phrenology 109, 140–45; and infant schools 174, 178; and Macnab 38

Penology: general interest of phrenologists in 145–6; and problem of capital punishment 151–2; psychological approach to 147–51
Phrenological Association, 88–9, 98 (see also Phrenologists, National Association of)
Phrenological Journal, 80–2
Phrenological Societies: activities 79; and Christian Phrenology 122; decline of 80, 91; distribution of 91; membership and organization of 40, 78–9
Phrenologists: agnosticism of 109; anti-metaphysical attitudes of 36, 169; anti-sectarianism 105–8; 113, 119–21; 133, 150, 173, 176, 185, 196; tendency to atheism 108–10, 128–30; on capital punishment 151; their collections 68–9, 94, 97; and corporal punishment 146, 150; as education reformers 190, 199, 228; enthusiasm for science 49, 85, 88, 119; estimated number of 91; and evolutionary theory 50–1; factionalism and jealousies of 93–95; health education and physiology 44, 73, 111, 186–7, 191; influenced by utilitarianism 127, 148, 152, 169, 171, 182; materialism of 39; National Association of 88–9, 98; and natural religion 104–12, 128–33, 172, 179; principles of education 167; popular scientists 58–9; popularity as lecturers 72, 85–7, 91; pro-German feelings 36, 69, 172, 181, 186, 220; reject classical education 170–1, 180–2, 190; reputation as 'Little Englanders' 69–71; and Sabbath Question 66–68; support of short hours campaign 66; views on Self Help 71
Phrenology: *a priori* judgments of 19; decline of 91–101; determinism of 22, 64, 87, 108, 116–7; discovery and delineation of faculties 21; and education 72, 127–8, 174; faculties in the brain 15–8, 21; fatalism of 63–4, 70, 116; on female minds 72; and

Index

free will 116; human progress and evolution 50, 68, 137–40, 144–5; impact in America 24–5; interaction of the faculties 16, 167; materialism of 65, 114–5, 124, 128, 136; and metaphysics 36, 60; nomenclature of 36, 122; on number of faculties 16, 20–1; optimism on British race 64, 68; other names for 12; opposed in the press 32, 39, 80 and physiology 44–5; popularity among women 72–3; and positivism 60; similarities with Owenism 140–1, 144–5; as a school subject 187–8, 211; as a universal science 19, 33; use of case-studies 14, 34–5, 39, 111, 120, 145, 188; use of classification 21, 35
Phreno-mesmerism, 47–8, 72, 95–99, 203
Physical education, 186, 191
Physiognomy, 13–4, 33, 35
Political economy, teaching of, 182–3
Prentice, Archibald, 155
Prince Albert, 172, 219–223
Proshaska, George, 13
Public debate and phrenology, 41, 42, 85, 91–2

Queen Victoria, 46, 220–21

Racialism, 68–71, 111, 123, 204
Rationalism, 8, 36–7, 70, 105, 110, 113–4, 167
Reid, Thomas, 36–9
Roget, Dr Peter, 34, 40, 42
Russell, Lord John, 29, 156, 200, 207–8, 218

Sampson, M.B., corresponds with Combe 90–1; his *Criminal Jurisprudence* 146; interest in homoeopathy 99; on necessity of work 65; on new sciences 101; on phreno-mesmerism 47
Scott, William, 19–20, 80, 117–8, 121, 132
Secularism: danger of 128; in education 179, 206–7, 211; in morality 68, 113; outlook on social reform 136, 141
Self-improvement, 22, 60–2, 68, 71, 218
Self-knowledge, 60–1, 64, 139
Simpson, James: and American education 173; on Australian conditions 153–4; on Bray's phrenology 87; critical of classical education 170–1; debate on education 196; on infant education 177; influenced by George Combe 191; on intemperance 62–3; leading lawyer 54–5; his natural religion 126, 132; on penal reform 161; on phreno-mesmerism 98; proposes system of education 174–8, 182; as reform-agitator 200; on self-reliance 65–6; sponsor of secular school 179; testimony before select committee 181, 196, 228
Smiles, R.W., 211, 213–6
Smiles, Samuel, 186
Smith, Joshua Toulmin, 61, 82, 96
Smith, Sidney, 84, 108, 139
Smith, William H., 140–1
Social virtues, 60–4, 71, 148, 183, 185
Spencer, Herbert, 21, 88, 96, 99
Spurzheim, Dr. J.C.: and British medical profession 48–9; debates Sir William Hamilton 41–2; on determinism 22, 116; and Deville 94–5; flexibility in phrenology 28, 39; and Gospel of Work 65; irreligion of 114–5, 132; parting with Gall 15; praised by medical journal 41; on propaganda tactics 77–8; as a public figure 12, 23–25; search for social psychology 15; visit to Cambridge 92
Stockmar, Baron Christian Friedrich, 220–2

Temperance: supported by phrenology 62–3, 68, 73; part of natural religion 111
Transportation System, 153, 155–6
Turner, Henry, 61, 68

Wallace, Alfred Russel, 71
Warne, Rev. Joseph, 121
Watson, Hewett C.: on crime and

punishment 146, 151, 159; democratic ideas 201, 205, 209–10; and *Phrenological Journal* 81–4, 100; and evolutionary theory 51–2, 54, 70; on local societies 78, 91, 96; love of politics 83; on phrenology's popularity 58, 86; and phreno-mesmerism 98; scepticism of 100, 109, 126, 228; and *Statistics of Phrenology* 23, 27, 40, 51

Welsh, Rev. David, 118–9, 121–2

Whately, Richard: on Australian conditions 153, 159; character of 221; a Christian Phrenologist 122–4, 132; and fringe sciences 99, 203; and political economy 183; and prison reform 146

Wilderspin, Samuel, criticized by phrenologists 177; and George Combe 191; and music 184; and variety of infant schools 175–8; 198

Williams, William H., education reformer 179–80; 184–5; religious indifference 179; and social sciences 183

Williams Secular School: attracts children of artisans 168; defended by phrenologists 178; non-sectarian education 179–80, 207, 211; phrenology in curriculum 188; supported by radicals 208; 217

Working classes: education and self-improvement of 64; factory petition 66; moral education of 168–9; separate phrenological society for 79; support of national education 212, 227; and Williams Secular School 168, 179

Wyse, Sir Thomas, 192, 196, 200–1

The *Zoist,* 81, 98–9, 166

For Product Safety Concerns and Information please contact our EU representative GPSR@taylorandfrancis.com
Taylor & Francis Verlag GmbH, Kaufingerstraße 24, 80331 München, Germany

www.ingramcontent.com/pod-product-compliance
Lightning Source LLC
Chambersburg PA
CBHW071822300426
44116CB00009B/1405